ELECTRON MICROSCOPY FOR BIOLOGISTS

ELECTRON MICROSCOPY FOR BIOLOGISTS

By

SASHI B. MOHANTY

B.V. Sc. & A.H., Ph.D.

Professor, Department of Veterinary Science
University of Maryland
College Park, Maryland

CHARLES C THOMAS • PUBLISHER

Springfield • Illinois • U.S.A.

Published and Distributed Throughout the World by
CHARLES C THOMAS · PUBLISHER
2600 South First Street
Springfield, Illinois, U.S.A. 62717

ISBN 0-398-04738-3
Library of Congress Catalog Card Number: 82-10286

With THOMAS BOOKS careful attention is given to all details of manufacturing and design. It is the Publisher's desire to present books that are satisfactory as to their physical qualities and artistic possibilities and appropriate for their particular use. THOMAS BOOKS will be true to those laws of quality that assure a good name and good will.

Library of Congress Cataloging in Publication Data

Mohanty, Sashi B.
Electron microscopy for biologists.

Bibliography: p.
Includes index.
1. Electron microscopy. I. Title.
QH212.E4M64 1982 578'.45 82-10286
ISBN 0-398-04738-3

Printed in the United States of America
PS-R-1

TO

MY MOTHER

PREFACE

This book is the outgrowth of a course in electron microscopy that I teach to graduate students in biologic sciences at the University of Maryland. In recommending a textbook to my students, I felt the need for a concise book that is primarily intended for biologists dealing with the basic principles of optics, instrumentation and operation of the electron microscope, basic and specialized techniques, and the latest developments in the field. The largest group of electron microscope users in the world is life scientists. Courses in biologic electron microscopy are available in many universities throughout the United States. There are several good textbooks on biologic techniques including some excellent series, optics, and the design of electron microscopes. However, there is a dearth of books that contain all this up to date information and the use of the electron microscope in a simple, understandable manner for the biologist.

Morphologic research has experienced an explosive growth due to the application of electron microscopy in the last two decades. All laboratories engaged in biologic or molecular biologic research possess or have access to an electron microscope. Many large hospitals have electron microscopes for diagnostic as well as research applications. The potential usefulness of electron microscopes in biology is immense, and there have been rapid advances in instrumental technology. Biologists are generally more familiar with descriptive rather than mathematical treatment of their work.

Electron Microscopy for Biologists is directed almost exclusively at the biomedical investigators who do not have a substantial background in physics and mathematics, and graduate students in electron microscopy. I hope that the book would also be useful to technicians, educators, and researchers not familiar with biologic techniques in electron microscopy. The book is based mainly on the experience gained from my teaching and working with the transmission electron microscope. It is divided into two parts: The

Transmission Electron Microscope and Techniques. Emphasis has been placed on the transmission electron microscope, the instrument most widely used in biology. Other related instruments, however, have not been neglected. The biologist requires a basic, simple understanding of electron optics and the principles of electron microscopy to appreciate the complicated instruments, their operational procedures, performance tests, problems encountered with them, and their potentials and limitations. In the first part, I have attempted to do so without much rigorous mathematical treatment, and the book is not limited to any particular current models of electron microscopes.

Electron microscopy, an applied science, is an art, and importance has been given to practical aspects of photography and specimen preparations in the second part of the book. This part, Techniques, contains all basic and commonly used specialized techniques including immunologic methods, autoradiography, and dark-field electron microscopy. The scanning electron microscope that has become quite popular among biologists has also been described along with the methodology for preparing biologic specimens. Importance has also been given to the high voltage and the most sophisticated scanning transmission electron microscopes, radiation damage, and image processing of electron micrographs. The use of the electron microscope in diagnosis of human diseases and the possibility of visualizing individual biomolecular atoms, which appears to be a distinct possibility, are described in the last chapter. Considering the enormous number of papers published on electron microscopy of biologic specimens, I have only provided selected references at the end of each chapter. The reader can refer to some of these papers and comprehensive reviews for full lists of references.

I am grateful to Dr. W. W. Marquardt, Department of Veterinary Science at this University, for reading the manuscript, making helpful suggestions and corrections. I am indebted to Mr. D. D. Rockemann, my research associate, for his assistance in photography. Sincere appreciation is expressed to Mr. M. E. Taylor, Jr., of the Central Electron Microscope Facility at this University for reviewing the chapter on scanning, scanning transmission electron microscopy and x-ray microanalysis, assistance in photography, and many helpful discussions. I am thankful to Dr. R. L. Steere of the Division of Plant Virology, U.S.D.A. and Mr. Gunther Thomas of the National Institutes of Health for many helpful suggestions. The secretarial assistance of Miss Margaret Kempf, who patiently typed the manuscript, made numerous changes and corrections, is especially appreciated. I thank Mr. Hwan Choi for the art work. Many scientists and electron microscope manufacturers from this country and abroad generously supplied many illustrations, which I have duly acknowledged in the legends. Thanks are due to my students, the faculty and staff of the Department of Veterinary Science at this University for their constant encouragement in writing this book.

Sincere gratitude is expressed to Charles C Thomas Publisher and their editorial staff for expert editing of the manuscript. Many helpful suggestions, encouragement, and patience of Mr. Payne E L Thomas greatly helped in this endeavor.

I will be gratified if this book would fulfill a need of those for whom it is intended.

S. B. MOHANTY

CONTENTS

PART I
THE TRANSMISSION ELECTRON MICROSCOPE

Chapter *Page*

PART II
TECHNIQUES

ELECTRON MICROSCOPY FOR BIOLOGISTS

PART I

THE TRANSMISSION ELECTRON MICROSCOPE

Chapter 1

BASIC PRINCIPLES OF LIGHT AND ELECTRON MICROSCOPY

MICROSCOPY IS THE ART and science of seeing very minute things. Fine structures of matter have always fascinated man, and he has made endless efforts to see them in increasing detail. The electron microscope (EM), built in the twentieth century, is the result of this pursuit to directly examine very small things that began in the seventeenth century with the construction of the light microscope (LM). Instruments involving vision, our most highly developed sense, have always been dominant in providing an insight into the structural organization of matter, and the modern sciences are overwhelmingly the fruit of vision.

While the optical or LM is simple in design and easy to use, the EM is a highly complicated, expensive, and inconvenient equipment. The EM, however, has the most important sole advantage over its counterpart. Its higher resolution can make clear and sharp images of objects 1,000× smaller than that of the LM. To the biologists, by resolving objects down to the large organic molecular level, it has opened up a new world, and it is the *key to the world of ultrastructure*. It has increased our understanding of the organization and structure of living things. The very concept of ultrastructure would have been impossible without this microscope. Indeed, this is the instrument that verified the theory of the existence of viruses. We have already seen some individual heavy atoms with it, and visualization of atoms of biomolecules remains a distinct possibility.

RESOLUTION, MAGNIFICATION, AND TYPES OF MICROSCOPES

Light refers to the electromagnetic radiation of those wavelengths or colors to which the human eye is able to respond directly. The LM refers to the microscope that uses light as the image forming radiation. The EM uses *electrons* for this purpose. *Resolution* (more accurately *resolving power* when referred to the instrument but not to the finished picture) is defined as the minimum separation between two objects at which they can be observed as separate entities and not as a blurred one. Qualitatively, it simply means the ability to discriminate fine details. The shorter the distance of separation, the higher or better the resolution. The unaided human eye can resolve two objects separated by a distance that subtends an angle of about 1 minute of arc (3×10^{-4} radian*) in front of the eye. It is capable of focusing at an infinite distance to about 25 cm (called *near point*, *normal working distance*, or *distance of comfortable vision*). At the near point, therefore, the resolution of the eye is 0.075 mm ($3\times10^{-4}\times25$ cm). Two objects separated by less than this distance will appear as one blurred object. Since this resolution requires an ideal contrast and is taxing on the eye to the limit of its performance, 0.2 mm is accepted as a more comfortable figure.

The resolution of an optical instrument is approximately one-half the wavelength of the image-forming radiation. Thus, the resolution limit of the LM that was reached in 1873 is about 2,000 Å.† To increase the resolution then, we need to use a form of radiation that has a shorter wavelength than light. The electromagnetic radiation spectrum consists of cosmic rays, gamma rays, x-rays, ultraviolet (UV) waves, visible light (wavelength = 4,000 Å to 8,000 Å), heat or infrared waves, radio, and electric waves. Only cosmic, gamma, x-rays, and UV waves have wavelengths shorter than light. Unfortunately, they cannot be used for image formation because the eye is insensitive to them. Waves shorter than UV cannot be refracted through a lens, and their radiation is lethal to many biologic specimens.

---

*1 radian = 57.3°

†Units of length measurements in electron microscopy:

1 mm = 1,000 micrometer (μm); 1 μm = 1,000 nanometer (nm)

1 nm = 10 angstroms (Å) (decinanometer); and 1 Å = 10^{-8} cm.

The wavelength of an electron, depending on the accelerating high voltage, is very small compared to light, e.g. 0.05Å at 50,000 volts v. 5,000Å of visible light. Since the wavelength of an electron is approximately 100,000× smaller than that of light, theoretically, the resolution of the EM should be better than the LM by this enormous factor. Unfortunately, due to the instrumental design and difficulties involved in forming electron images, this has not been achieved in practice. However, the microscope has achieved 1,000× better resolution than its counterpart, which is indeed remarkable. It also can directly magnify 500× more than the LM. A simple way to resolve fine structure of an object is to magnify it so that the eye can appreciate the greater details of the magnified image. How much magnification does one need? This is determined by calculating the *maximum useful magnification* (MU), which is the ratio of the resolution of the eye to the resolution of the instrument. For the LM, this magnification is $\frac{0.2 \text{ mm}}{2{,}000 \text{ Å}} = 1{,}000\times$. The MU for an electron microscope with a resolution of 2Å is 1,000,000× or 1,000× greater than the light microscope. Any magnification in excess of this figure reveals no further detail. In fact, the picture then becomes blurry and deteriorates in quality. This excess magnification is called *empty magnification.*

Of the three types of electron microscopes—conventional transmission electron microscope (TEM), scanning electron microscope (SEM), and the newest, most sophisticated scanning transmission electron microscope (STEM), TEMs are in maximum use. The largest group of TEM users in the world today is life scientists. Compared to LMs, the TEMs appear sufficiently complex to prospective users so that an attempt to understand their operation usually precedes their use. The basic optical principles of both the instruments are virtually the same. A biologist needs a fundamental understanding of the TEM for its proper use, specimen preparation, and interpretation of electron micrographs. Also required is a knowledge of the basic principles of light optics.

Historical Synopsis

Microscopy had its beginning when Leeuwenhoek in Holland saw *animalcules* (little animals) with his simple microscope in 1675. Robert Hook, from London, is credited with building the first compound LM in 1677. Then it took almost 200 years of development before the LM reached its peak state of perfection. This is probably because the instrument was mainly used by people for whom the microscope was a hobby. They had very little knowledge of the basic principles of optics, and it was easy to see some tiny things without understanding the instrument. Instead of improving the optical design of the instrument, they were interested in decorating their microscopes with fine leather inlaid with gold and silver filigree. Toward the middle of the nineteenth century, Zeiss, in Germany, recognized that these people had reached the limit of their knowledge and that further improvement of the LM needed the help of physicists and mathematicians. He approached Abbé, the German mathematician, for this purpose. Based on the work done by Airy and Rayleigh in 1873, he formulated a relationship between the wavelength of light, the aperture opening of the microscope, and the medium through which light waves pass. Abbé's equation showed that the resolution of the LM is limited by the nature of light beyond which no further improvement is possible. From the very beginning, physicists, mathematicians, and engineers have been involved in constructing and improving the TEM. Thus, this microscope achieved a remarkably comparable level of LM after only twenty years.

Research into properties of cathode rays started in the 1850s by Giessler, and in 1897 Thomson showed that electrons are negatively charged particles that could be deflected by electromagnetic or electrostatic fields. In 1924, de Broglie hypothesized a wave nature for electrons and calculated their wavelength. In 1926, Busch demonstrated that a suitable electromagnetic or electrostatic field could focus electrons and act as a lens. It remained for Knoll and his student Ruska in Germany to demonstrate the first TEM in 1931. It had a magnification of 17× with a two-stage lens system, but the resolution was much poorer than that of the LM. Rüdenberg, who contributed nothing toward the development of the EM, applied for a patent in Germany in 1931. In German patent laws, he is erroneously called the inventor of this microscope.

In the same year, Brüche and Johannson built the first electron emission EM with a resolution of 10 μm. Naturally, these results brought a lot of discouragement (which caused Knoll to leave for the field of television), and the instrument was never expected to outperform the LM. However, Ruska, in 1933, single-handedly built a true electromagnetic TEM that surpassed the optical microscope in resolution and magnification for the first time (Fig. 1-1). The first electron micrograph of a biologic specimen, a section of a root, was taken in Brussels by Marton in 1934. In 1935, Knoll demonstrated the theory of the SEM, and the first SEM was built by von Ardenne in 1938. A resolution of 100 Å was achieved by von Borries and Ruska in 1938, and the commercial production of TEM by Seimens in Germany began in 1939. An electrostatic TEM was first built by Boersch and Mahl in 1939. The commercial production of TEM in North America was begun by Radio Corporation of America in 1941, and a resolution of 10Å was achieved by Hillier in 1946.

In the early 1930s, TEMs were built from some parts available in flea markets after World War I. The first commercial SEM was built by Cambridge Instrument Company in 1965. In 1970, Crewe and coworkers built a STEM with a field emission source. At first,

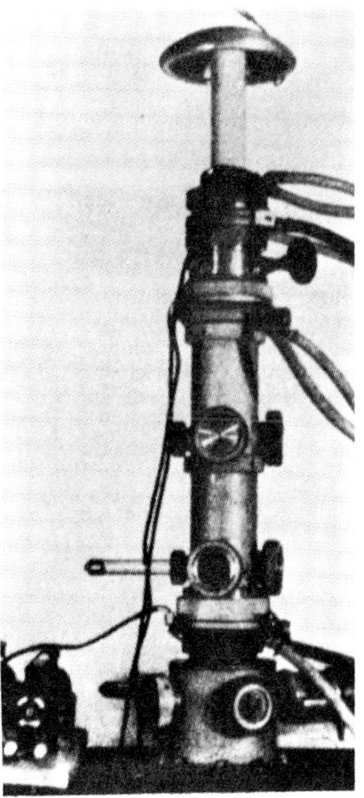

Figure 1-1. The first transmission electron microscope (Ruska, 1933), that surpassed the resolving power of the light microscope.

biologists were pessimistic about the usefulness of TEM. Because of the high voltage in the electron gun, some people were afraid of the instrument. However, the TEM soon became an invaluable tool in biology. Biologists, the largest group using the TEM, usually demand simplicity and high performance from their instruments. The present-day TEMs, therefore, have undergone a revolution in their design and performance.

NATURE OF LIGHT

Of the four theories of light (corpuscular, wave, electromagnetic, and quantum theories), the wave nature of light satisfactorily explains many basic principles of optics. Accordingly, light travels outwards in concentric spherical waves from the source. This wave motion resembles water waves produced by throwing a pebble into the water. The water waves move outwards from the center (where the pebble was dropped), forming circles of increasing radius that become parallel at infinity. It is generally difficult to explain optical phenomena with waves and wavefronts. Instead, the motion of wave fronts may conveniently be indicated by rays, each travelling perpendicular to the wavefront (Fig. 1-2). In a transverse representa-

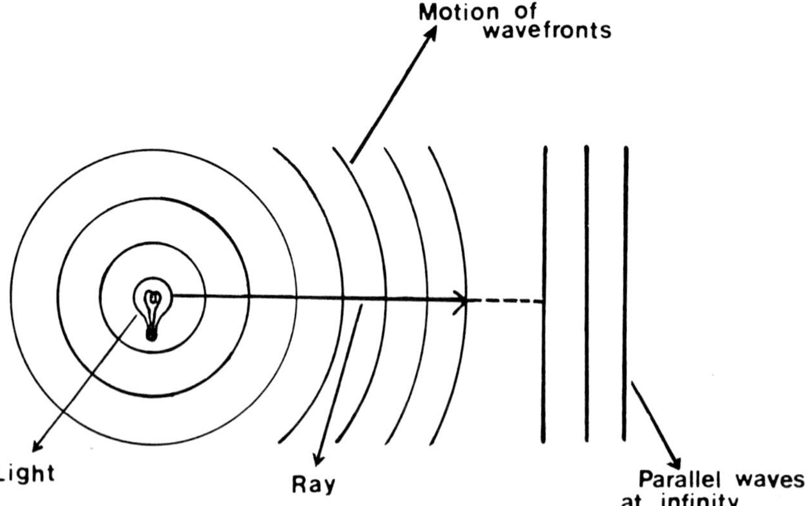

Figure 1-2. Wave motion of light. Adapted from G.A. Meek, *Practical Electron Microscopy for Biologists*, 2nd ed. Copyright © 1976. Reprinted by permission of John Wiley & Sons, Ltd., London.

tion of the waves (sine waves), the characteristics are defined by *wavelength* (λ), which is the distance from one wave crest to the next (Fig. 1-3), and the *amplitude* (crest-to-trough height of the waves). Thus, the λ of the visible light varies from 4,000 to 8,000Å (violet to red). The time relationship or phase of light waves causes *interference* of light. If two identical waves are out of phase (Fig. 1-3), they will cancel each other when superimposed, and a dark band will result (destructive interference). However, if the two identical waves arrive at the same place at the same time exactly in phase, they will reinforce, and a bright band of light appears (constructive interference). The wave nature of light also produces diffraction, which is considered in Chapter 3.

Glass Lenses, Geometrical Optics, and Ray Diagrams

Biconvex glass lenses, which may be considered to be made up of a large number of sections of prisms, are used to focus light rays by bending or refracting them from their original path in a predictable manner. Refraction of light occurs as it passes between media of two refractive indices. A light *ray* is considered as an infinitely thin pencil

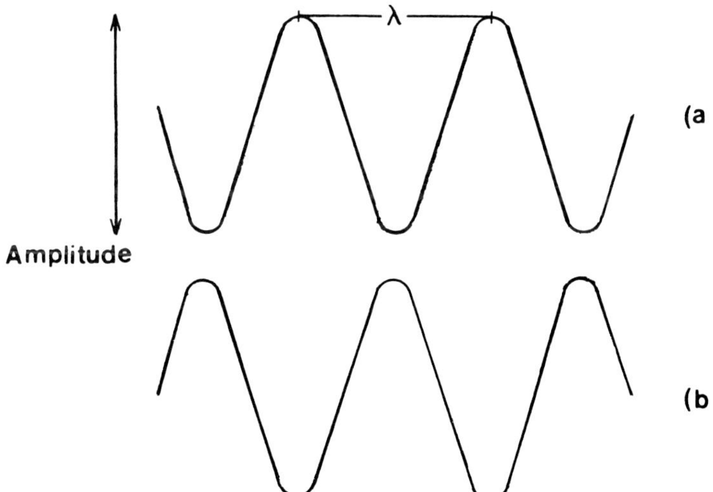

Figure 1-3. Transverse representation and phases of light waves. Waves a and b are exactly out of phase and will cancel out each other when superimposed. Adapted from G.A. Meek, *Practical Electron Microscopy for Biologists*, 2nd ed. Copyright © 1976. Reprinted by permission of John Wiley & Sons, Ltd., London.

or beam (which actually is not possible due to diffraction effect, *see* Chap. 3). The focusing action is due to a change in the velocity (slowing down) of light during its passage through the glass lens. When a beam of parallel rays strikes a thin biconvex lens in a direction perpendicular to the central plane (Fig. 1-4), the rays, after passing through the lens, will converge to a focal point (F_2) in the image space (back focal point) and then diverge again. A *focal point* is defined as the point to which rays from an object situated at an infinite distance will be converged. The lens also has a focal point (F_1) in the object space (front focal point). The distance from the focal point to the center of the lens is the *focal length* (f). The shorter the focal length, the stronger the lens. The straight line passing through the center of the lens and the focal points is called the *axis* or *principal axis*.

The space to the left of the lens is the *front*; to the right is the *back*. Parallel rays from a source of light placed at an infinity distance will reveal a *real image* of the source on a screen placed behind the lens in the focal plane. This real image plane of an object at infinity is called the *principal focal plane* or *Gauss plane*. A lens of this type is called a *positive* or *convergent* lens. This lens can also focus nonaxial light rays even if they do not strike the lens perpendicularly. A biconcave lens that has a diverging action is called a *negative* lens. Both positive and negative lenses are used in the LMs, but the electromag-

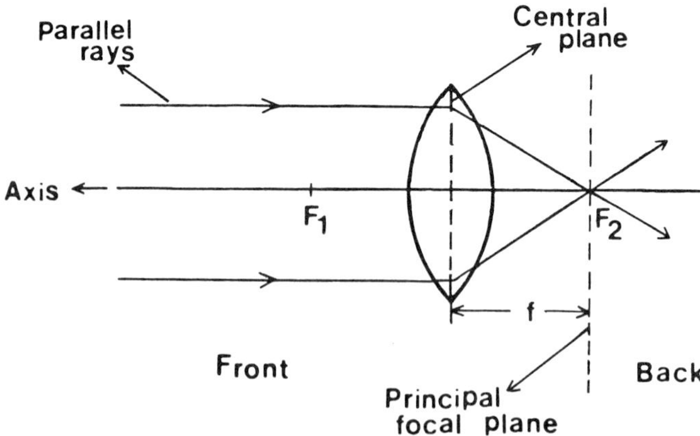

Figure 1-4. Ray diagram through a biconvex glass lens. F_1 and F_2 are front and back focal points, respectively; f = focal length.

netic lenses of EMs are always positive. These lenses cannot be made negative; electrostatic ones can be (*see* Chap. 2).

Geometrical optics are based on ray diagrams, that assume that the lens is extremely thin, rays parallel to the axis are focused on the axis, and the rays passing through the geometrical center of the lens are not deflected and pass through straight regardless of the direction from which they come. The *geometrical center* or *center* of the lens is the point where the axis intersects its central plane. In a diagram with a horizontal axis, the rays are assumed to travel from left to right; from top to bottom in a vertical axis.

REAL IMAGES: The real image formed by a convergent lens can be determined by the ray diagram in Figure 1-5a. Let P, an upright arrow, be an object in front of the lens OO situated at a distance slightly greater than the focal length of the lens. A ray drawn from P parallel to the axis, after traversing the lens, is bent through the back focal point F'_0. A ray from P through the center of the lens passes through undeviated. The point of intersection of these two rays locates P′, the position of the image P. Rays from each point of the object P can be traced in the image plane in this manner. The image P′ is a *real image* because it can be projected on a screen. This image is also magnified and inverted (note the direction of the arrow). Thus, the lens OO forms a real, magnified, and inverted image of the object P situated at a distance slightly greater than the focal length of the lens. The *object distance*, p, in front of the lens, is the distance from the object to the center of the lens. The *image distance*, q, behind the lens, is the distance from the image to the center of the lens. The fundamental lens formulas are as follows:

$$\frac{1}{p} + \frac{1}{q} = \frac{1}{f}, \qquad \text{and } M \text{ (\textit{Magnification})} = \frac{q}{p}$$

The magnification will depend on the object distance, p. Where $p = q = 2f$, $M = 1$, and the image is of the same size as the object. If $p > 2f$, then $M < 1$, and the image is smaller than the object. If the object is situated between $2f$ and f, $M > 1$, and the lens magnifies.

VIRTUAL IMAGES: Let us consider the image formation with another lens EE that is larger than the lens OO (Fig. 1-5b). The object P′ (inverted arrow) is situated in front of this lens, but at a distance slightly less than its focal length. If we draw the two rays

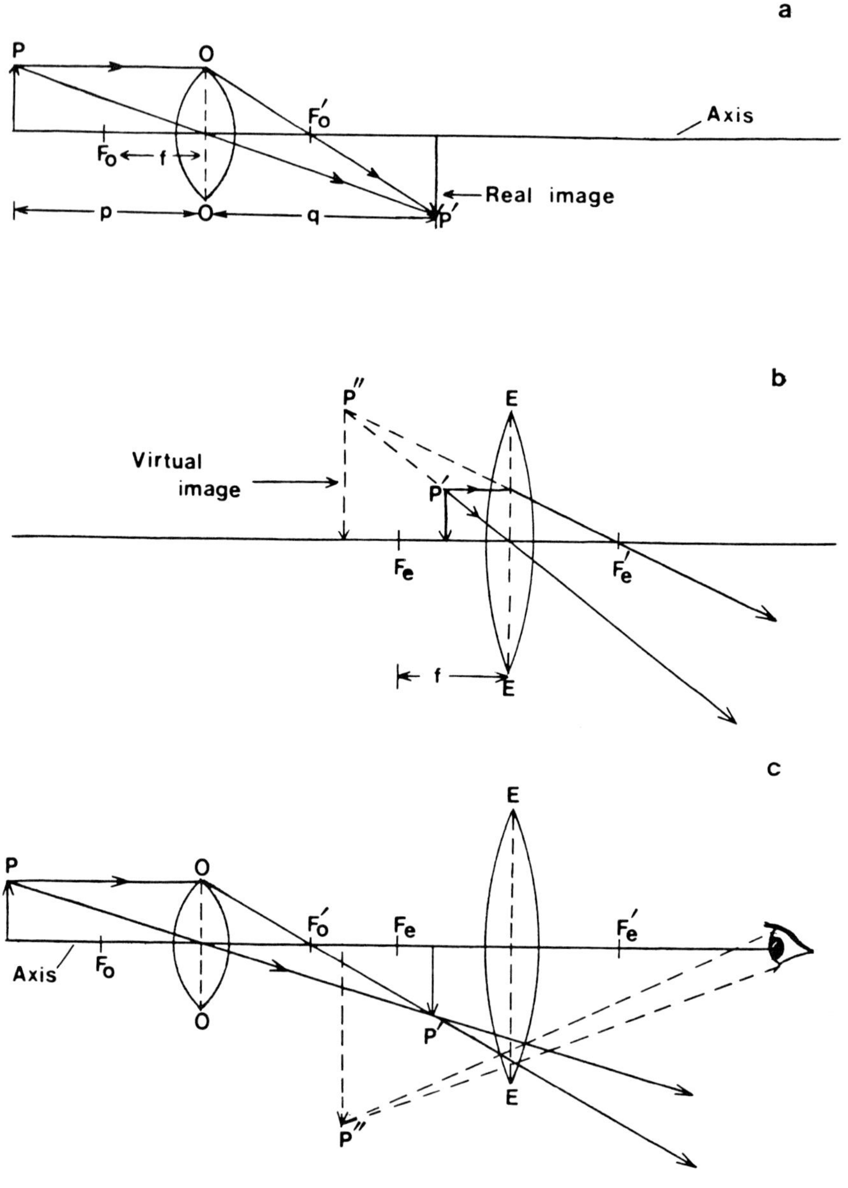

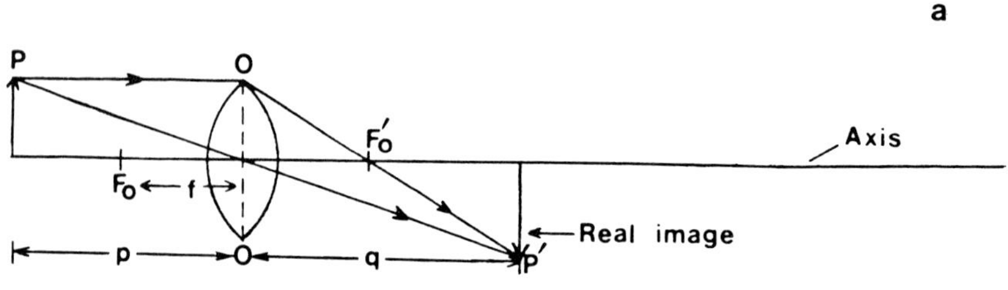

Figure 1-5. Ray diagrams showing image formation in two different glass lenses and in a compound light microscope. (a) Real image formation by the objective lens. (b) Virtual image formation by the eye piece. (c) Image formation in a compound light microscope (combination of Figs. 1-2a and b).

from P′, these rays will pass through the lens. The parallel ray is refracted through the back focal point, F_e'; the ray through the center passes undeviated. However, after leaving the back of the lens, these rays will diverge, and the lens cannot form a real image. Another lens system capable of forming a real image, or the optical system of the eye placed behind this lens, however, can make the divergent rays converge and form a real image. The rays forming the image P″ appear to come from P″ and they do not actually pass through this point. This apparent image is magnified but not inverted (note the direction of arrow); it is in front of the lens (unlike the image formed by lens OO, Fig. 1-5a), and it cannot be made visible by projecting on a screen. It can only be seen by another optical system such as the eye. It is, therefore, called a *virtual image.* The intermediate lens that controls the magnification is used in this manner in some TEMs. It is apparent that when $p < f$, the image is virtual.

IMAGE FORMATION IN THE COMPOUND LIGHT MICROSCOPE: This microscope may be considered to consist of two thin simple convex lenses, one OO (Fig. 1-5a) acting as the objective, the other EE (Fig. 1-5b) as the eyepiece (ocular). The ray diagram in the microscope (Fig. 1-5c) can be obtained by combining Figures 1-5a and b. The objective lens OO forms a real, magnified, and inverted image, P′, called the *primary image.* The eyepiece EE further magnifies P′ and a *secondary* virtual image, P″ is formed. The two rays emerging from the lens EE appear to diverge from P″ and cut the axis in the region slightly beyond the focal point of the eyepiece. An eye placed here would focus the divergent beam to form a real magnified image of P on the retina. This can be applied to all points on the object P so that a complete image is built upon the retina. The total magnification is the product of the magnifications of the two lenses.

It is apparent from Figure 1-5c that the eyepiece has to be rather large to collect all the rays coming from the object P. The objective usually has a shorter focal length (strong lens); the eyepiece has a large one (weak lens). Actually, the path of rays in a compound microscope is much more complicated than that shown here. However, our ray diagrams through very thin lenses capable of forming point images of point sources are helpful in understanding the principles of all optical instruments. They are also fairly accurate for illustrating how the TEM works.

ANGULAR APERTURE AND APERTURE ANGLE: The *angular aperture*

(Fig. 1-6) is the angle of the cone of light accepted by a lens. The larger this acceptance angle (2α), the more information the lens can gather. Aperture angle (α) is one-half of the angle of the cone of light accepted by the lens.

The Function of a Microscope

The eye is an imperfect optical system because the retina is made of rod and cone cells and its resolution is limited. Furthermore, its angular aperture is very small—a maximum angle of 8° subtended by the pupil at its near point (25 cm). An oil immersion objective has a feasible angular aperture of approximately 175°. A microscope is, therefore, used to magnify the image and collect more information concerning the object so that the eye can resolve it clearly.

Nature of Electrons and Electron Emission

An electron is one of the basic constituents of the atom. It is a small, negatively charged particle with a mass of 9×10^{-31} kg (9×10^{-28} gm), a diameter of 10^{-12} cm, and a charge of 1.6×10^{-19} coulomb. It travels in a straight line, is propagated in waves, has energy, and produces fluorescence after striking a fluorescent screen. For a simple understanding, the atomic nucleus is surrounded by an atmosphere of electrons that are moving around it in elliptical orbits (Fig. 1-7). The atomic nucleus is the dense, positively charged central core made of protons and neutrons. The number of protons in the nucleus is called the atomic number, Z, and the number of electrons increases with increasing atomic numbers. A strong

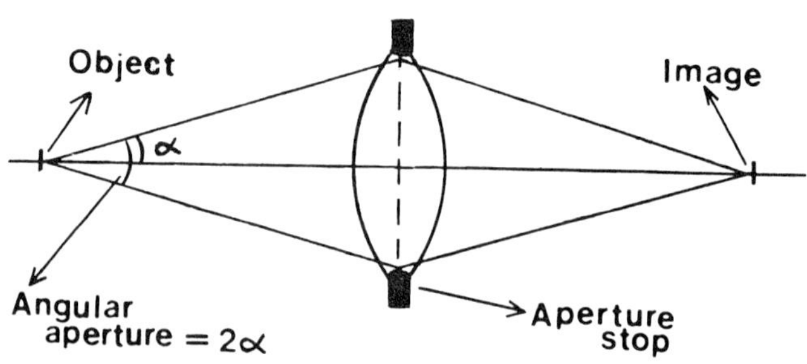

Figure 1-6. Angular aperture of a glass lens. Aperture angle (α) is one-half of the cone of light accepted by the lens.

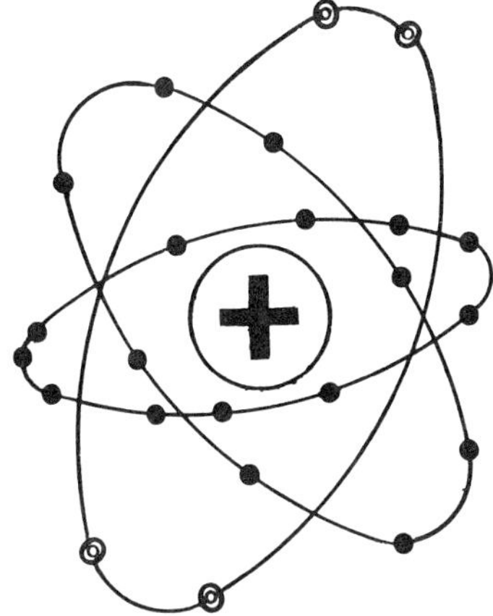

Figure 1-7. A simple representation of atomic nucleus and electrons around it. Adapted from S. Wischnitzer, *Introduction to Electron Microscopy*, 3rd ed. Copyright © 1981. Reprinted with permission from Pergamon Press, Ltd., New York.

attracting electric field of the positively charged nucleus prevents the electrons from escaping from the orbit. However, this force is least for valence electrons travelling in the outer orbit because it is largely balanced by the neutralizing negative charge of the electrons in the inner shells. Thus, the valence electrons can be more readily detached than the rest from the atom.

In a metal, the attracting force of neighboring atomic cores or *ions* causes the valence electrons to be easily detached. They are called *free electrons* and are free to move (Fig. 1-8). A change in their velocity and direction of motion occurs when they collide with each other and with ions. Some electrons near the metal surface, therefore, may escape. However, this leaves a net positive charge on the metal surface, at the point from which the electron came. The attractive force tends to pull the electron back to the metal again. The electron must spend energy to overcome this positive force. The magnitude of the energy required by the electron to escape the surface is called *work function*. The necessary energy is provided by

Figure 1-8. Positive ions and free electrons on a metal surface. Adapted from S. Wischnitzer, *Introduction to Electron Microscopy*, 3rd ed. Copyright © 1981. Reprinted with permission from Pergamon Press, Ltd., New York.

heating the metal by applying an electric current. This is called *thermoionic emission* and is used in most electron microscopes. Electrons can also be emitted by applying light energy (photoemission) and strong electric force (field emission). Field and secondary emissions are discussed in Chapter 17.

Wavelength of an Electron

The λ of an electron according to de Broglie's equation $= \frac{h}{mv}$ where in mksc (meter, kg, seconds, and coulomb) system, h = Planck's constant $= 6.6 \times 10^{-34}$ joules/sec; m = mass of electron $= 9 \times 10^{-31}$ kg; and v = velocity of electron.

From the energy relationship, $KE = {}^1\!/_2 mv^2 = Ve$; where V = voltage, e = charge on electron $= 1.6 \times 10^{-19}$ coulomb.

If we substitute the value of v in de Broglie's formula,

$$\lambda \text{ of an electron} = \frac{h}{m\sqrt{\frac{2Ve}{m}}} = \frac{h}{\sqrt{2Vem}} \quad \text{or} \quad \lambda = \sqrt{\frac{h^2}{2Vem}}$$

by substituting all the values;

$$\lambda \text{ of an electron} = \sqrt{\frac{(6.6 \times 10^{-34})^2}{2 \times (9 \times 10^{-31}) \times V \times 1.6 \times 10^{-19}}}\ \text{m}$$

$$\sqrt{\frac{150 \times 10^{-20}}{V}}\ \text{m} = \frac{12.3 \times 10^{-10}}{\sqrt{V}}\ \text{m} = \frac{12.3 \times 10^{-8}}{\sqrt{V}}\ \text{cm} = \frac{12.3}{\sqrt{V}}\ \text{Å}$$

Therefore, the λ of an electron is inversely proportional to the square root of the high voltage applied to accelerate the electrons.

Analogy Between Light and Transmission Electron Microscopes

Since the two microscopes use image-forming radiations that differ greatly in their nature, the construction and operation of the TEM is quite different from the compound LM. The LM may be considered as an inverted microscope for comparison with the TEM (Fig. 1-9). The source of illumination is an electric light bulb (or sunlight) in the LM, but it is a *V*-shaped tungsten filament in the TEM. Glass and electromagnetic lenses are used in LM and TEM, respectively. The light passes through air, glass, and oil, but the TEM needs a high vacuum (10^{-4} to 10^{-6} torr (mm Hg)) for the electrons to travel in the microscope column. The LM is focused by moving the objective lens in respect to the specimen to judge the sharpness of the image. The magnetic field strength of the objective lens is altered for focusing electron images. The specimen for LM usually rests on a glass slide (1 to 2 mm thick), but the specimen support film has to be very thin (50 to 200Å) for transmitting the electrons. The image can be observed with the eye and recorded on photographic plates in both microscopes. A fluorescent screen is provided for viewing electron images.

SELECTED BIBLIOGRAPHY

Agar, A.W., Alderson, R.H., and Chescoe, D.: *Principles and Practice of Electron Microscope Operation.* In Glauert, A. (Ed.): *Practical Methods in Electron Microscopy*, vol. 2. Amsterdam, North-Holland, 1974.

Barer, R.: *Lecture Notes on the Use of the Microscope.* Oxford, Blackwell, 1953.

Fisher, R.B.: *Applied Electron Microscopy.* Bloomington, Indiana University Press, 1954.

Hall, C.E.: *Introduction to Electron Microscopy*, 2nd ed. New York, McGraw-Hill, 1966.

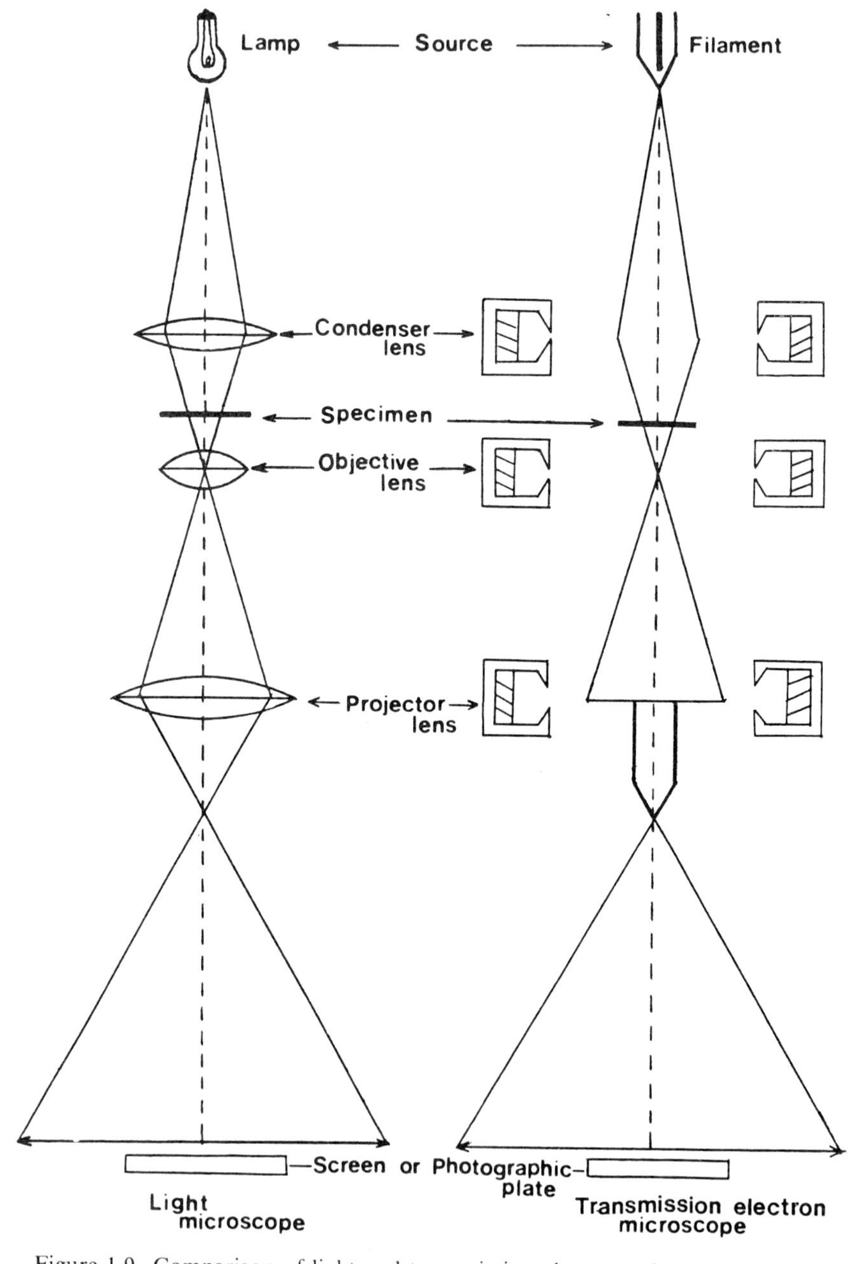

Figure 1-9. Comparison of light and transmission electron microscopes.

Meek, G.A.: *Practical Electron Microscopy for Biologists*, 2nd ed. London, Wiley, 1976.
Mulvey, T.: The development of electron optics. *Proc EMSA, 29th Annual Meeting*, 2, 1971.
Sjöstrand, F.S.: *Electron Microscopy of Cells and Tissues*, Vol. 1. *Instrumentation and Techniques*. New York, Academic Press, 1967.
Wischnitzer, S.: *Introduction to Electron Microscopy*, 3rd ed. New York, Pergamon, 1981.
Wyckoff, R.W.G.: *The World of Electron Microscope*. New Haven, Yale University Press, 1954.
Zworikin, V.K., Morton, G.A., Ramberg, E.G., Hillier, J., and Vance, E.E.: *Electron Optics and the Electron Microscope*. New York, Wiley, 1948.

Chapter 2

ELECTRON OPTICS

ELECTRON OPTICS deals with the propagation of electrons as light optics deals with that of light. Busch, in 1926, demonstrated that an electromagnetic or electrostatic field had properties of a lens. Lenses for electron microscopy can be electromagnetic, electrostatic, or a combination of both. Electromagnetic lenses have many advantages over their counterparts, and they are commonly used in electron microscopes (EMs).

MAGNETIC FIELD

Unlike light, an electron beam does not change its velocity during its passage through a magnetic field, its refracting medium. Furthermore, glass lenses have sharp boundaries; light passes abruptly from a medium of one refractive index (air or immersion fluid) to glass and then abruptly to air again. Electrons are continuously refracted in a magnetic field, and there are no sharp boundaries between the magnetic field and the immersion medium (vacuum in the microscope). The magnetic field around a permanent magnet is represented by *lines of force* running from the north to the south pole (Fig. 2-1). The total number of lines of force represents the *magnetic flux*; the number of lines of flux per unit area is the *flux density*. The relation between magnetism and electricity was demonstrated by Oersted in 1819. He found that a compass needle, when placed near a conducting wire carrying an electric current, is deflected. A straight conductor carrying a current creates a magnetic field around the conductor, which attracts iron filings. The magnetic field of this conductor is illustrated by a series of concentric circles,

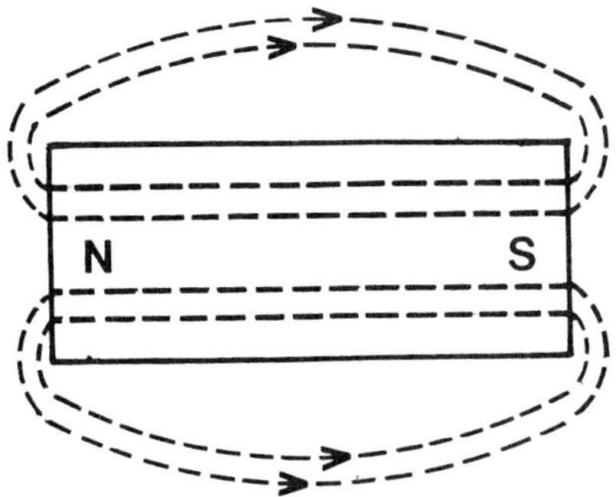

Figure 2-1. Lines of force in a bar magnet. Adapted from S. Wischnitzer, *Introduction to Electron Microscopy*, 3rd ed. Copyright © 1981. Reprinted with permission from Pergamon Press, Ltd., New York.

the lines of force (Fig. 2-2). If the conductor is a loop, the lines of force circle around the wire (Fig. 2-3), and the flux density is maximal at the center of the loop. If several loops of wire are wound around a cylindrical surface to make a *coil* or *solenoid*, each turn contributes to the magnetic field generated (Fig. 2-4). The flux density at the center of the solenoid is

$$B = \mu \left(\frac{NI}{l}\right)$$

where B = flux density; μ = permeability; N = number of turns; I = strength of the current; and l = length of the solenoid.

Since magnetic field strength $H = \frac{NI}{l}$, then, $B = \mu H$, and $\mu = \frac{B}{H}$ = (flux density per unit field strength).

If the core of the solenoid is air or other nonconducting material, the solenoid does not generate a strong electromagnetic field because the μ of air and nonmagnetic materials is 1. A soft iron core with high μ (several thousands), however, creates an intense magnetic field, and strong electromagnets usually have a ferromagnetic core. Electromagnets are basic components of magnetic lenses of most EMs. Since the magnetic field strength of electromagnets can be changed at will by varying the current passing through these magnets, they are more versatile than permanent magnets. The focal length, f, of an electromagnetic lens is

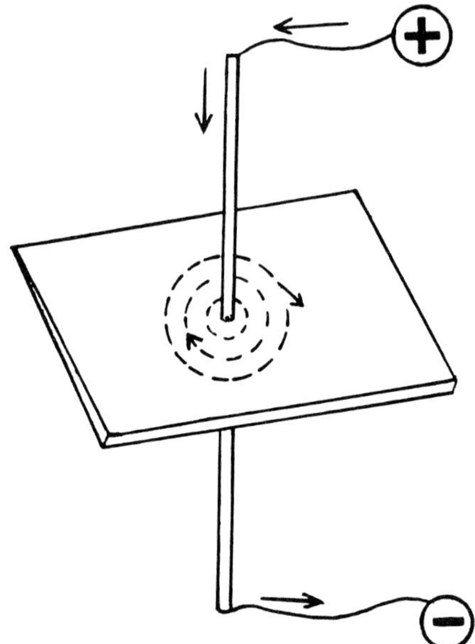

Figure 2-2. Magnetic field and lines of force around a conducting wire. Adapted from S. Wischnitzer, *Introduction to Electron Microscopy*, 3rd ed. Copyright © 1981. Reprinted with permission from Pergamon Press, Ltd., New York.

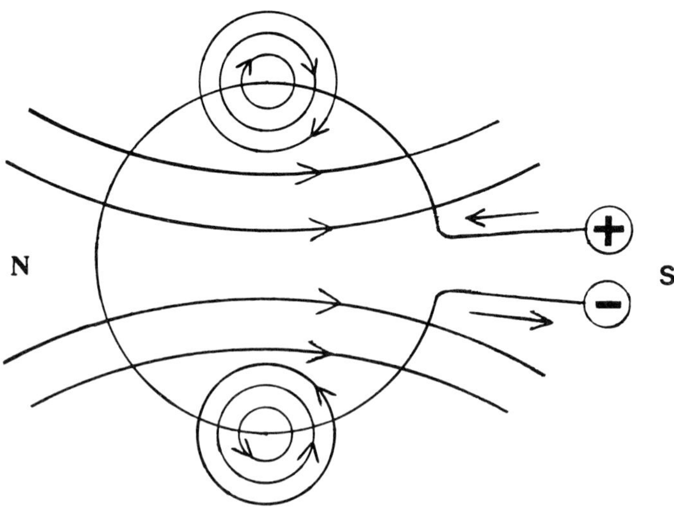

Figure 2-3. Lines of force around a conducting loop. Adapted from F.S. Sjöstrand, *Electron Microscopy of Cells and Tissues*, Vol. 1, *Instrumentation and Techniques*, 1967. Courtesy of Academic Press, New York.

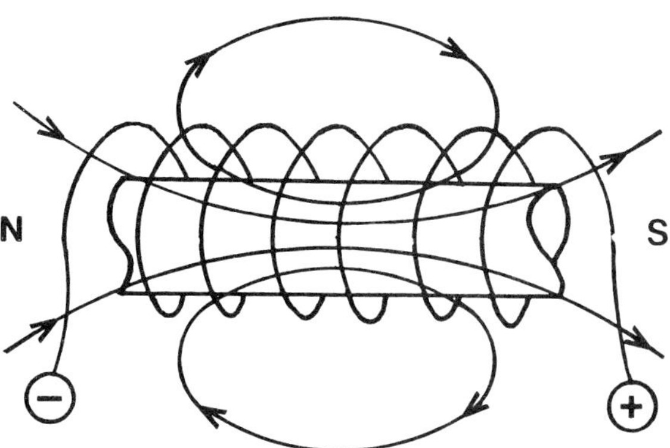

Figure 2-4. Lines of force in a solenoid. Adapted from F.S. Sjöstrand, *Electron Microscopy of Cells and Tissues*, Vol. 1, *Instrumentation and Techniques*, 1967. Courtesy of Academic Press, New York.

$$f = k\left(\frac{V}{NI^2}\right)$$

where k = a constant; V = accelerating voltage; N = number of turns in the coil; and I = lens current. As the current in the coil is increased, the lens becomes stronger and the f decreases. The lens becomes weaker and the f increases as the current flowing through the coil is decreased.

Relationship of Lens Strength with Current

Magnetization of a soft iron (hence lens strength) varies with the lens current (magnetizing force) applied, and their relationship is shown in Figure 2-5. When a nonmagnetized lens starts from the point O, the lens strength increases (shown by the sigmoid path OA) as the lens current is increased. The lens becomes saturated at point A, and no further increase in lens strength occurs. When the lens current is reduced, the lens strength does not retrace the original path AO; rather it is displaced and follows the path AB. Even after the field strength is reduced to zero at point B, some magnetism still remains in the soft iron. This is called remnance. The phenomenon of magnetization lagging behind the force that induces magnetism (current) is called *hysteresis*. A reverse current OC must be applied to remove the residual magnetism (bring back to zero).

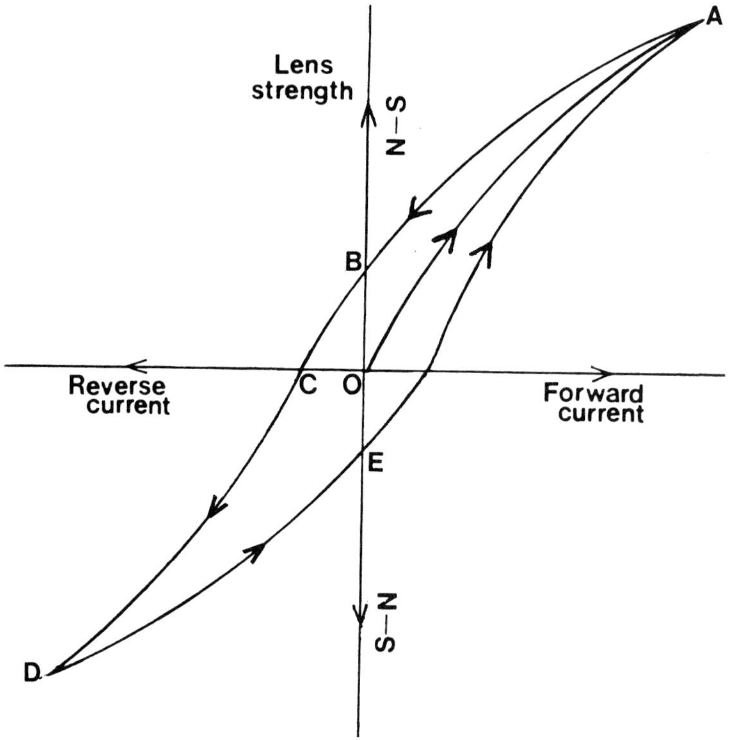

Figure 2-5. The hysteresis loop. Adapted from G.A. Meek, *Practical Electron Microscopy for Biologists*, 2nd ed. Copyright © 1976. Reprinted by permission of John Wiley & Sons, Ltd., London.

The lens strength will increase again as the reverse current is increased. The lens will be saturated at the point D, and the polarity of the lens will be reversed. This change in polarity does not change the focusing action of the lens, but the spiral path of electrons inside the lens is reversed. When the reverse current in the lens is reduced, the lens strength follows the curve DE and again passes through the remnance point. A forward current applied to induce magnetism would then increase the lens strength in the direction EA. When the direction of the magnetic field is changed back and forth, the curve illustrating the relationship between lens strength and current is a closed loop, called the *hysteresis loop*. Due to hysteresis, lens strength cannot be exactly calculated from the current flowing in it.

Hysteresis can cause ±10% error in magnification.

When a conductor carrying a current enters a magnetic field in a perpendicular direction, it is subjected to a force that acts on it in an opposite direction. This makes the conductor move in a direction at right angles to the direction of the field as well as at right angles to the direction of the current. This was demonstrated in a Faraday motor or dynamo. Fleming's *left-hand-thumb* or *motor rule* will apply to the conductor as it passes through the magnetic field (Fig. 2-6). The index finger of the left hand in this orientation points in the direction of the magnetic field (B). The middle, ring, and little fingers together indicate the direction of the current (I), and the thumb points in the direction of force (F). An electron beam is also subjected to this force as it passes through a magnetic field. However, electrons flow from the cathode to the anode (opposite to

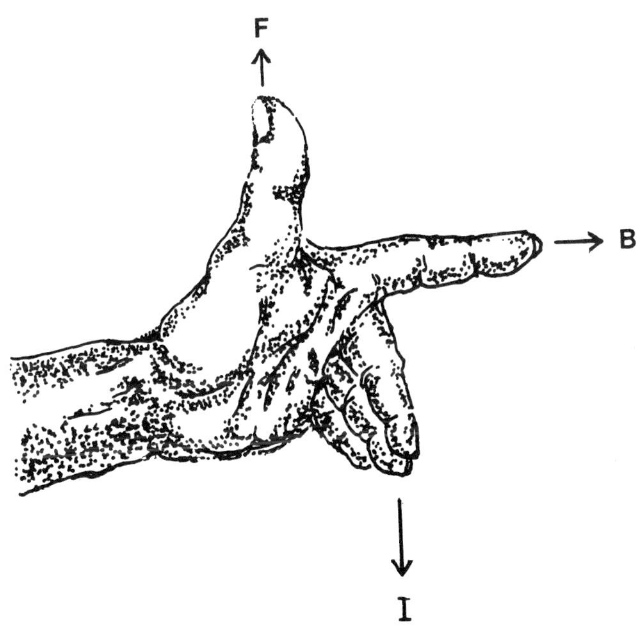

Figure 2-6. Fleming's left-hand-thumb or motor rule. F = force; B = direction of magnetic field; and I = direction of current. Adapted from F.S. Sjöstrand, *Electron Microscopy of Cells and Tissues*, Vol. 1, *Instrumentation and Techniques*, 1967. Courtesy of Academic Press, New York.

the direction of current), and the right-hand-thumb rule is applicable to them.

Therefore, when an electron enters a large uniform magnetic field in a perpendicular direction, the force acting on it is directed both at right angles to the direction of the flow of the electron and the direction of the magnetic field. As the electron continues to move through the magnetic field, the diverting force acts on it the whole time it is within the field and causes the electron to assume a circular path by the time it exits from the lens (Fig. 2-7). However, an electron (or conductor) moving parallel to the magnetic field is not affected by the force, and it is not deviated from its path.

A cone of electrons from an object point usually enters the magnetic field at an acute angle in the EM. The path of the electron through this field can be explained by vector components. Since an electron has velocity, it falls under vector quantity, and its vector, V, can be resolved into two vectors, V_1 and V_2 (Fig. 2-8a). The velocity component V_1 acting at right angles to the field makes the electron inscribe a complete circle (Fig. 2-8b). The velocity component V_2 (acting parallel to the field), is not affected, and the electron moves straight through. A simultaneous action of these two velocity components, therefore, makes the electrons take a helical (spiral) path (Fig. 2-9). A number of electrons with the same velocity from an object point will be focused after a 360° rotation at an image point on the axis. This action is basically similar to that of a

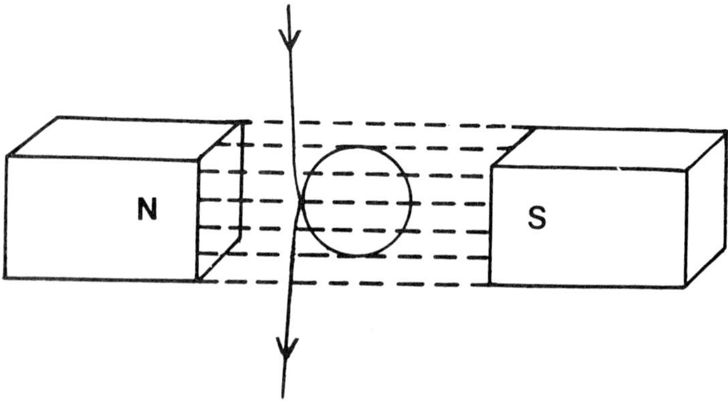

Figure 2-7. Circular path of an electron through a magnetic field entering it perpendicularly.

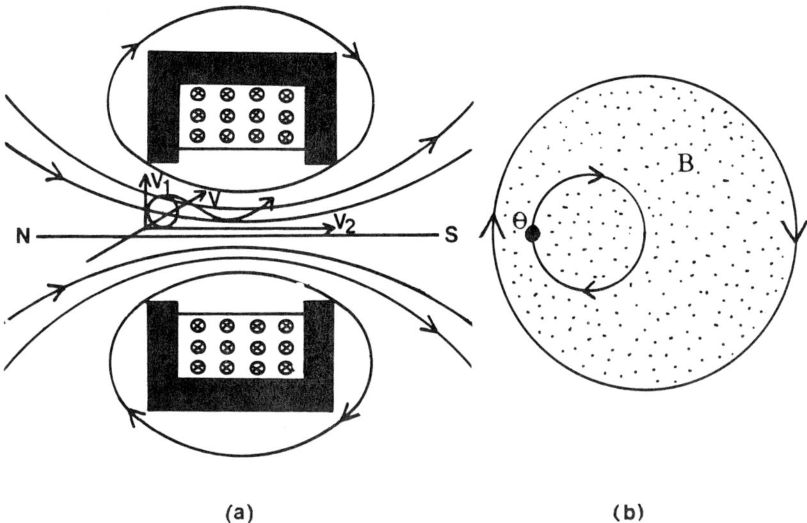

Figure 2-8. Vector components of an electron in an electromagnetic field. (a) An electron moving from point 0 with a velocity V. Its vector V can be resolved into V_1 and V_2. Adapted from F.S. Sjöstrand, *Electron Microscopy of Cells and Tissues*, Vol. 1, *Instrumentation and Techniques*, 1967. Courtesy of Academic Press, New York. (b) Circular path of Vector V_1 in cross section.

convergent glass lens. Thus, the ray diagrams and lens formulas for light optics are also applicable to electron optics.

Design of Electromagnetic Lenses

The magnetic lenses for the EM must be strong and have a short focal length. The magnetic field in the center of a simple solenoid is very weak because the field is spread out for a long distance along the axis. An iron shroud (encasing) can be used to shield the coil, and the magnetic field is then concentrated at the center of the coil (Fig. 2-10a). If the shielding is extended further by covering the inner surface of the coil except for a small ring-shaped gap, the field is concentrated at the magnetic gap (Fig. 2-10b). A still more concentrated field needed for constructing a lens of short focal length can be obtained by adding a soft iron polepiece at the annular gap (Fig. 2-10c). Polepieces were introduced by von Borries and Ruska in 1932, and they provide a highly concentrated magnetic

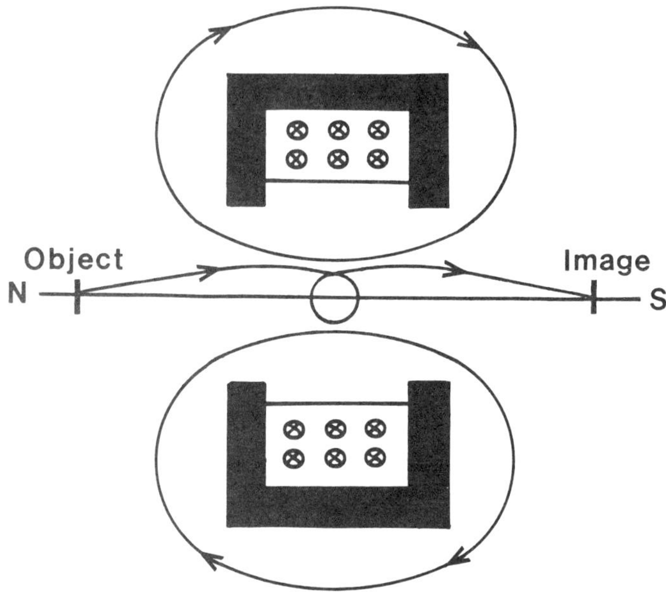

Figure 2-9. Helical path of an electron passing through an electromagnetic field.

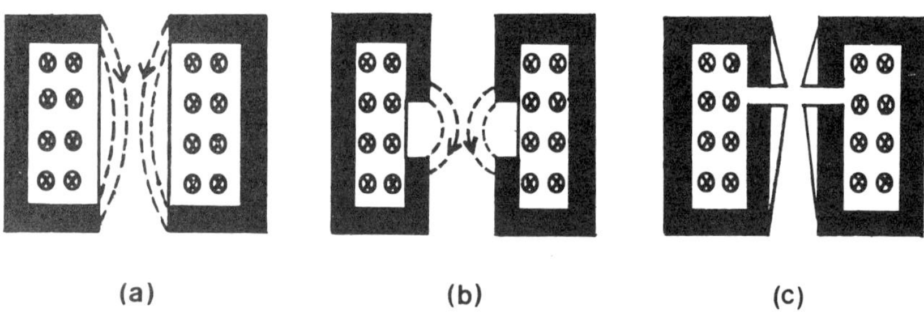

Figure 2-10. Construction of electromagnetic lenses. (a) A shrouded lens. (b) Encasing of the lens except at a small gap. (c) A magnetic lens with a polepiece.

field over a very short distance (a few mm) along the axis. Polepieces are made of two cylindrical pieces of iron with a nonmagnetic brass or copper spacer and have a small bore of only a few mm (Fig. 2-11).

Polepieces are used to achieve a very small focal length (up to 1 mm) for a high resolution, high magnification, for reducing the physical size of magnetic lenses, and to prevent overheating. They are standard parts of most magnetic lenses used in the EM.

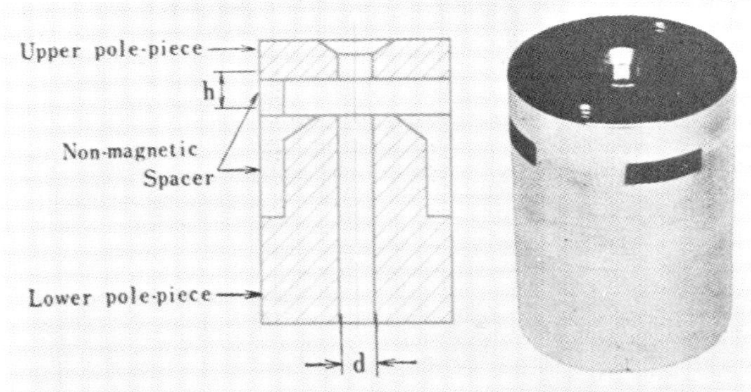

Figure 2-11. A polepiece (on the right) and its longitudinal section (on the left) showing the polepiece components. Courtesy of Hitachi, Ltd.

However, polepieces can not be machined perfectly, and due to inhomogenities of polepiece material (iron), they produce asymmetrical magnetic fields. This asymmetry is the principal source of astigmatism in electron microscopy (*see* Chap. 3). The magnetic field strengths generated by lenses of different design are shown in Figure 2-12.

PROPERTIES OF ELECTROMAGNETIC LENSES: These lenses deflect electrons without changing their velocity, all magnetic lenses are positive, and their strength (hence focal lengths) can be varied by controlling the current passing through them. They also have hysteresis and aberrations (*see* Chap. 3), and the image rotates in relation to the object.

FOCUSING OF ELECTROMAGNETIC LENSES: The focal length of the objective glass lens in a light microscope is fixed. The focusing is done by judging the sharpness of the image by moving the lens toward or away from the object. The positions of magnetic lenses and the specimen are usually fixed in an EM. Therefore, focusing is achieved by varying the objective lens current in the microscope.

Electrostatic Lenses

Although electrostatic fields can also focus an electron beam, their demerits outweigh their merits over magnetic lenses. Therefore, electrostatic lenses are not commonly used in EMs. However, the electron gun in the illuminating system of all EMs is a lens of this type (*see* Chap. 5).

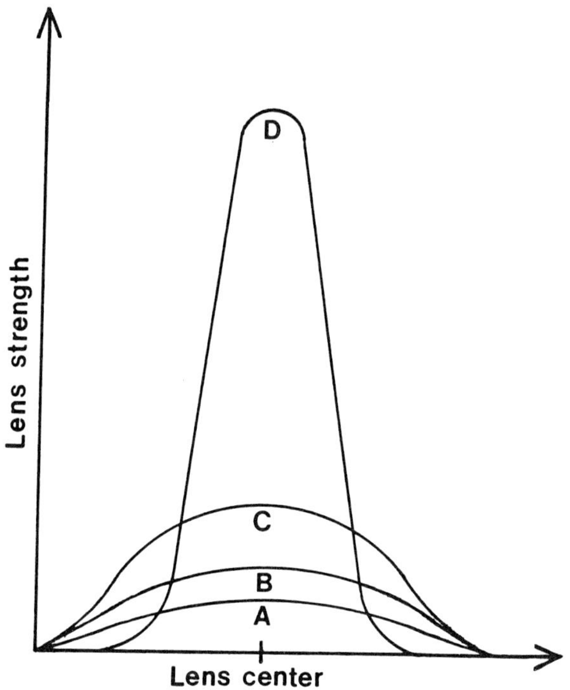

Figure 2-12. Magnetic field strength distribution curves. (a) Field generated by a simple solenoid. (b) A shrouded lens. (c) A lens shrouded throughout except a small gap. (d) A polepiece lens.

A simple electrostatic lens can be built by connecting two metal plates to a battery and by applying a potential difference (in volts) between them. Electrons would then be attracted by the positive plate. The *lines of force* in the electric field are perpendicular to the plate, and a uniform electrostatic field can be illustrated by *equipotential surfaces* that are parallel to the plates (perpendicular to lines of forces). A series of spherically curved equipotential surfaces can be established by inserting a charged plate with a central opening between the two plates maintained at different potentials. The curved equipotential surface would then act as a lens. When the potential difference between the central plate and the electrodes is greater on one side of the central plate, the equipotential surface would bulge into the space in which the field is weaker (Fig.

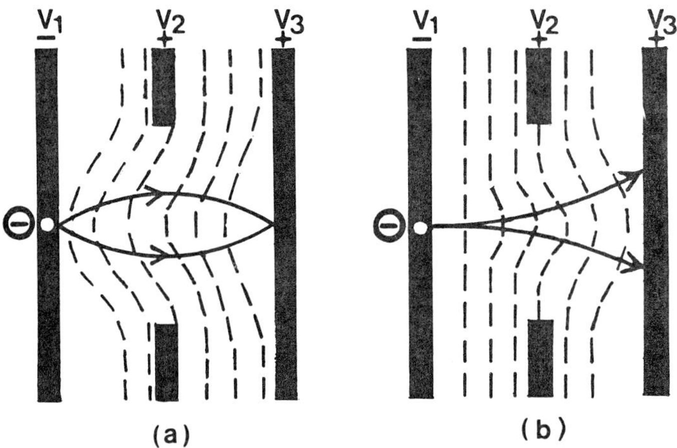

Figure 2-13. Action of electrostatic lenses. (a) Positive lens, $V_2 - V_1 < V_3 - V_2$. (b) Negative lens, $V_2 - V_1 > V_3 - V_2$. Adapted from F.S. Sjöstrand, *Electron Microscopy of Cells and Tissues*, Vol. 1, *Instrumentation and Techniques*, 1967. Courtesy of Academic Press, New York.

2-13). Electrons entering the field in a direction perpendicular to the plates will converge (positive lens) if the equipotential surface toward the electrons is convex (Fig. 2-13a). They will diverge (negative lens) if the equipotential surface is concave toward the electrons (Fig. 2-13b).

The focusing action of an electrostatic lens is similar to that of a glass lens. An electric field causes a change in electron velocity (as in light) and direction as it passes through the field. The electrons pass through a gradually increasing field strength as they travel toward the center of the lens. Then they encounter a decreasing electrostatic field strength until they exit the lens. Unlike the glass lens, however, there is no sharp boundary to the electrostatic field.

The advantages of electrostatic lenses over electromagnetic lenses are that (1) the voltage for the lenses and accelerating high voltage are derived from the same source and a voltage divider always maintains a constant ratio between the two voltages. Therefore, the accelerating potential can be changed without a change in magnification or focus (opposite as occurs in electromagnetic lenses); (2) correction of astigmatism is easier, and since there is no hysteresis, field strength can be calculated exactly; and (3) the image remains in its original orientation because it does not rotate, and a small size instrument is possible.

The disadvantages of these lenses are that (1) their focal length cannot be made very short, and the spherical aberration and chromatic aberration constants for these lenses are very high. Therefore, the resolution is poor; (2) the lenses need high potential, which can cause electrical breakdown (arching over), particularly when the vacuum is low. They have limited accelerating potential up to 50 KV; and (3) distortion of field occurs due to contamination deposits that become charged by electron bombardment.

Focusing of electrostatic lenses is usually done mechanically. There are only a few commercially built electrostatic transmission EMs.

SELECTED BIBLIOGRAPHY

Cosslett, V.E.: *Practical Electron Microscopy*. New York, Academic Press, 1952.

Fisher, R.B.: *Applied Electron Microscopy*. Bloomington, Indiana University Press, 1954.

Grivet, P.: *Electron Optics*, Oxford, Pergamon, 1965.

Haine, M.E.: *The Electron Microscope*. London, Spon, 1961.

Hall, C.E.: *Introduction to Electron Microscopy*, 2nd ed. New York, McGraw-Hill, 1966.

Meek, G.A.: *Practical Electron Microscopy for Biologists*, 2nd ed. London, Wiley, 1976.

Siegel, E.M. (Ed.): *Modern Developments in Electron Microscopy*. New York, Academic Press, 1964.

Sjöstrand, F.S.: *Electron Microscopy of Cells and Tissues*, Vol. 1, *Instrumentation and Techniques*. New York, Academic Press, 1967.

Wischnitzer, S.: *Introduction to Electron Microscopy*, 3rd ed. New York, Pergamon, 1981.

Zworikin, V.K., Morton, G.A., Ramberg, E.G., Hillier, J., and Vance, E.E.: *Electron Optics and the Electron Microscope*, New York, Wiley, 1948.

Chapter 3

ABERRATIONS AFFECTING MICROSCOPE PERFORMANCE

To yield a faithful image, a microscope must have a high resolution, it must have a large depth of field, and the object must produce sufficient contrast so that the observer can study the image characteristics. We will consider these factors in both the light and electron microscopes.

RESOLUTION

Various factors limit the resolution of a microscope. Of these, diffraction is the single most important factor in limiting the performance of the light microscope (LM).

Diffraction and Airy Disk

Diffraction is defined as a modification in distribution of waves when a wavefront strikes a barrier. This interference in the path of wavefronts bends the waves and gives rise to secondary wavefronts at the edge. We have considered wave motion of light in Chapter 1. Diffraction of light can be easily explained by analogy with diffraction of water waves. Let us place a barrier with a central opening across the middle of a small water pool (Fig. 3-1). If we throw a pebble into the pool, water waves will start from the point where the pebble hits the water and will travel outwards. All these waves will be reflected back except at the opening, where the wavefronts will bend around the corner and create secondary wavefronts.

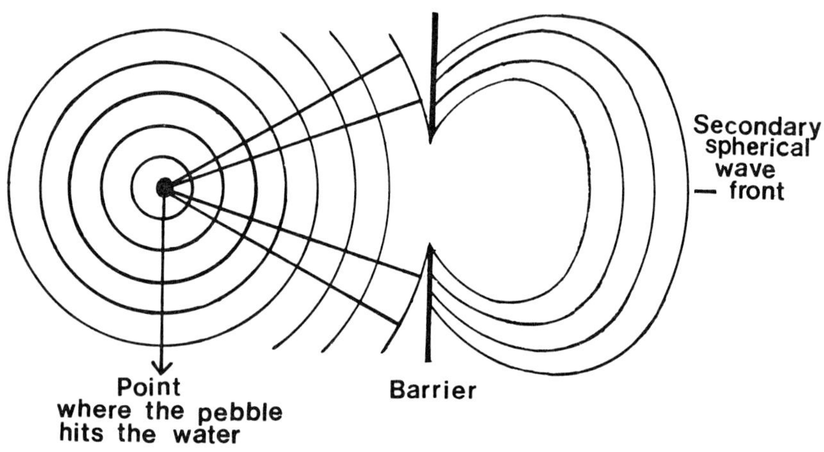

Figure 3-1. Diffraction of water waves. Adapted from S. Wischnitzer, *Introduction to Electron Microscopy*, 3rd ed. Copyright © 1981. Reprinted with permission from Pergamon Press, Ltd., New York.

If a perfect point source, e.g. a pinhole in a screen, is illuminated by an intense beam of light and the light passing through the hole is focused through an apertured biconvex lens on another screen (Fig. 3-2), the image formed is not a pinpoint of light. Instead, it consists of a central mass of light surrounded by several diffuse rings. The pattern produced is called an "Airy disk," named after the discoverer. The central bright disk contains 84% of light and the remainder is spread into surrounding rings. A diffuse Airy disk will be produced even if a perfect lens and a monochromatic light source were used. The edge of the limiting aperture (aperture stop) is responsible for the diffraction effect. This defect cannot be eliminated by removing the limiting aperture because opaque areas of a specimen and the edge of a lens would serve to produce diffraction. Therefore, diffraction is an inherent characteristic of wave motion, in general that limits our ability to see a perfect image.

Aperture stop is a circular hole placed at or near the center of the lens. It can also be placed at the front or back focal plane of the lens.

Limit of Resolution

Airy, Rayleigh, and Abbé derived an equation for the diffraction

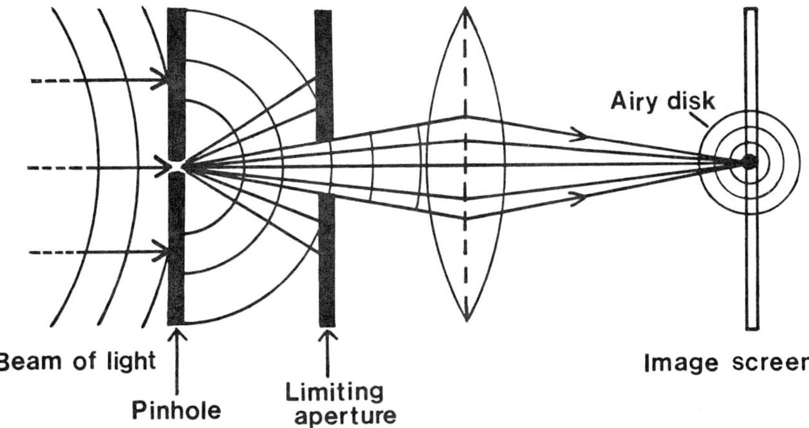

Figure 3-2. Diffraction of light waves and Airy disk. Adapted from S. Wischnitzer, *Introduction to Electron Microscopy*, 3rd ed. Copyright © 1981. Reprinted with permission from Pergamon Press, Ltd., New York.

effect. It is now known as the Abbé's equation and it is as follows:

$$d = \frac{0.61\lambda}{n \sin \alpha}$$

where d is the radius of the first dark ring; λ is the wavelength of light; n is the refractive index of the medium between the lens and the object; and α is the objective aperture angle (one-half of the angular aperture). The value of the constant 0.61 is somewhat controversial and varies from 0.61 to 1.22.

This expression was originally derived from astronomical work and strictly applies to infinitely small self-luminous objects like stars. However, it is also applicable to objects examined in a LM that are illuminated externally, provided the aperture angle of illumination falling on the specimen is approximately equal to the aperture angle of the lens (Fig. 3-3). Abbé realized that a lens with a large α would collect more information on the object. He introduced the concept of *numerical aperture* (N.A.) of a lens, the light-gathering power of the lens, as follows: N.A. (always marked on light microscope objectives) = $n \sin \alpha$.

The significance of Abbé's equation, expressing the magnitude of diffraction effect, is that it determines the resolution of an aberration-free LM. This can be explained by considering two small,

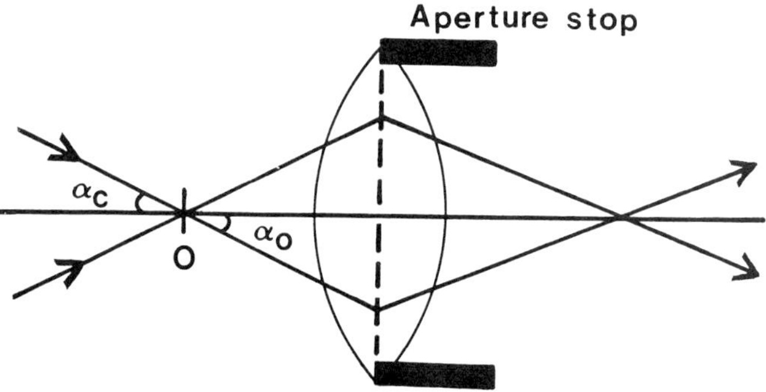

Figure 3-3. The path of light from an object illuminated externally. α_c = aperture angle of illumination reaching the specimen plane; α_o = aperture angle of light traveling from the object plane to the lens (aperture angle of lens); and O = object plane. Adapted from S. Wischnitzer, *Introduction to Electron Microscopy*, 3rd ed. Copyright © 1981. Reprinted with permission from Pergamon Press, Ltd., New York.

equally bright self-luminous objects separated by a short distance. When imaged through a glass lens, each will produce an Airy disk and, depending on the distance of separation between the objects, the disk will overlap (Fig. 3-4). If the disks do not overlap appreciably, i.e. they are separated by the diameter of the central disks, the two objects are fully resolved. The intensity profile of an

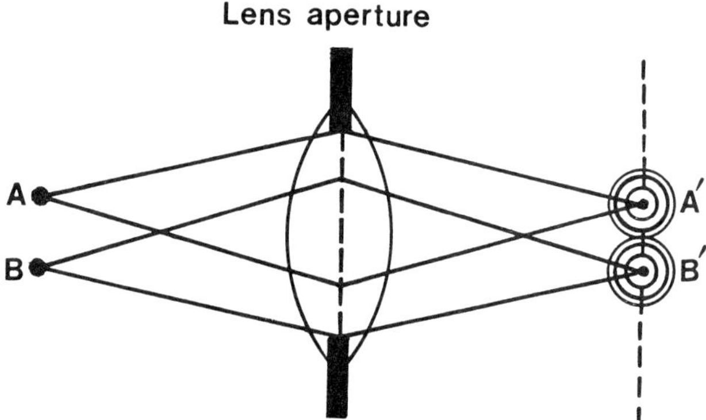

Figure 3-4. Resolution of two point objects. The objects A and B give rise to Airy disks A′ and B′ respectively. Adapted from S. Wischnitzer, *Introduction to Electron Microscopy*, 3rd ed. Copyright © 1981. Reprinted with permission from Pergamon Press, Ltd., New York.

Airy disk is shown in Figure 3-5a, and the intensity pattern of two disks overlapping each other is illustrated in Figure 3-5b. Rayleigh proposed that there is a minimum distance of separation between the center of the two Airy disks for possible resolution. This is equal to the radius of the first dark ring. At this distance, there is an intensity drop of 19% between the two bright peaks, which is somewhat arbitrary, but practical. If the two peaks approach more closely, it is impossible to say if the images are of two separate objects, or of an oblong object.

Figure 3-6 represents an actual photograph of Airy disks. The

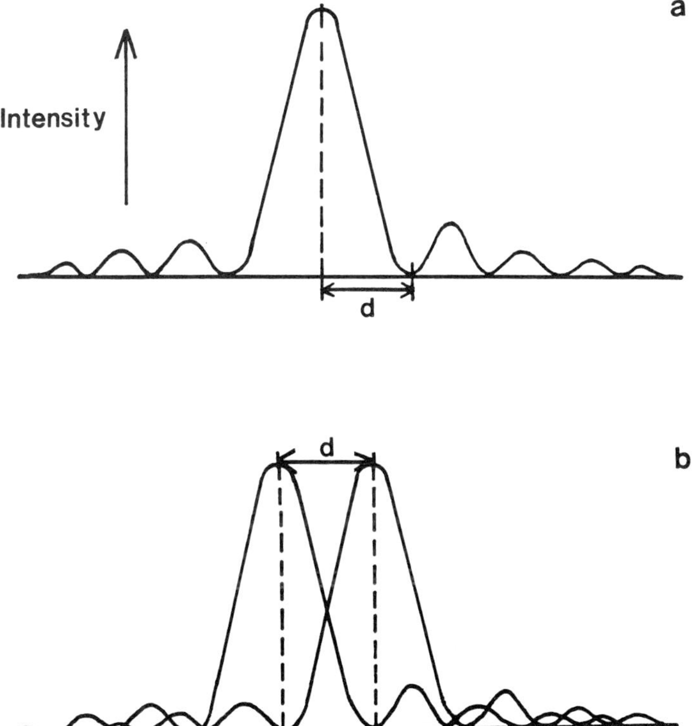

Figure 3-5. Intensity profile of Airy disks. (a) Microdensitometer tracing of an Airy disk. The central peak contains 84% of light and the other contains the remainder. (b) Rayleigh's criterion of resolution in terms of separation of two Airy disks. The central maximum intensity of one image coincides with the first minimum intensity of the other; d = minimum distance between the center of the central peak and the center of the first dark fringe.

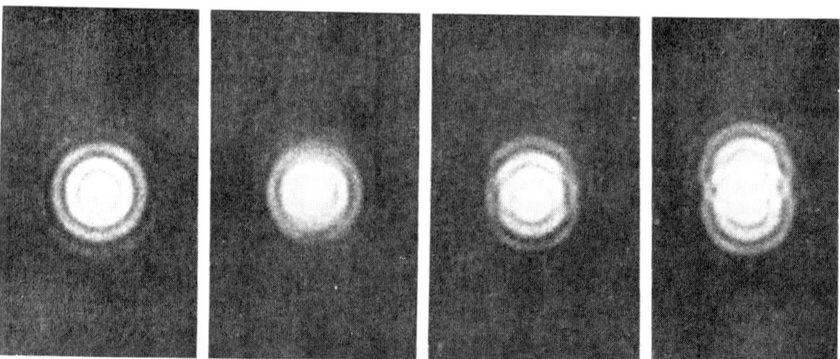

Figure 3-6. Photograph of Airy disks. (left to right) (a) Airy disk arising from one point object. (b) Disks of two point objects are separated by one-half the radius of the central disks. (c) Pattern produced when the distance of separation is equal to the radius of the disks. (d) Disks formed when the distance of separation equals twice the radius of the central disks. From *Principles of Optics* by A.C. Hardy and F.H. Perrin. Copyright © 1932. Used with the permission of McGraw-Hill Book Company.

disk arising from one point object is seen in Figure 3-6a. When the two points are separated by one-half the radius of the central disks, the disks appear as one (Fig. 3-6b). The pattern produced by two point objects separated by a distance equal to the radius of the disks (d) is seen in Figure 3-6c. When the distance of separation equals twice the radius of the central disks, the objects are resolved (Fig. 3-6d). It is apparent that a minimum distance of d is required for possible resolution.

We can then calculate the resolution of the LM due to diffraction effects as follows:

$$d = \frac{0.61\lambda}{n \sin \alpha}$$

where λ of visible white light = 5,000Å.

n = 1.5 for most immersion oils (approximately the same as glass); α = 70° (an ideal angle should be 90°). This is not possible because the specimen has to be placed away from the lens); and $\sin \alpha = 0.94$

By substituting the values

$$d = \frac{0.6 \times 5{,}000}{1.5 \times 0.94} = 2{,}163\text{Å} \simeq 0.2\ \mu\text{m}$$

This is the best resolution possible in a LM due to the diffraction, the only important factor that limits the resolution of the microscope. As Abbé pointed out, the resolution is limited by the wavelength of light (which should be as short as possible) and N.A. (which should be as large as possible). For a better resolution, then, we need a radiation of shorter wavelength.

The diffraction of electrons is similar to that of the light, and Abbé's equation can be used to calculate the resolution of an electron microscope due to diffraction effects as follows:

The λ of an electron derived by de Broglie's equation (Chap. 1)

$$= \frac{12.3}{\sqrt{V}} \text{ Å}$$

Thus, we can determine the resolution of a transmission electron microscope (TEM) with an accelerating potential of 50 KV and with α of 10^{-2} radians. Compared to the LM, the aperture angle of the objective lens is extremely small. Therefore, $\sin \alpha \simeq \alpha$. The refractive index, n, of vacuum is 1, so $n \sin \alpha$ (N.A.) $= 1 \times .01 = .01$.

Thus, the λ of electrons in this microscope

$$= \frac{12.3}{\sqrt{50{,}000}} = 0.05\text{Å}.$$

By substituting the values

$$d = \frac{0.61 \times 0.05}{0.01} = 3\text{Å}.$$

Since the λ of an electron is extremely small, the effect of diffraction on the resolution of a TEM is minimal.

Fresnel Diffraction

Diffraction at a sharp or opaque edge is known as *Fresnel diffraction*. The effect makes the edge appear blurred due to the formation of a series of fringes called *Fresnel fringes*, named after the investigator. The mechanism of formation of these fringes with light is explained in Figure 3-7. Fringes are formed when secondary waves created at the edge interfere with the primary wavefronts.

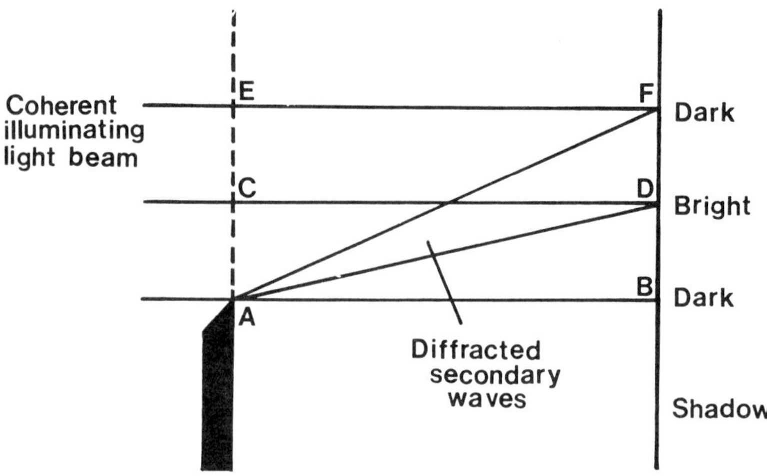

Figure 3-7. Mechanism of Fresnel fringe formation with light. Diffraction at a straight edge, which is illuminated by coherent light waves, gives rise to secondary waves, which interferes with the primary wavefronts. If the diffracted wave AD differs from primary wave by one full wavelength, a bright fringe is formed due to reinforcement of waves. If they differ by one-half wavelength as in AF, they are cancelled and a dark fringe results. Adapted from G.A. Meek, *Practical Electron Microscopy for Biologists*, 2nd ed. Copyright © 1976. Reprinted by permission of John Wiley & Sons, Ltd., London.

Fresnel Fringes in the Electron Microscope

When a beam of electrons strikes an opaque edge, e.g. the edge of a hole in a carbon support film, or a sharp object, Fresnel fringes are produced. This is caused by interference between scattered and unscattered electrons at the edge. The characteristic changes at the edge can be observed at high magnification as the objective focusing is altered. The strong first order single bright Fresnel fringe is normally seen in the electron microscope. However, with a coherent electron source using a very small aperture, fully defocused condenser system, and a long exposure (3 to 5 minutes), a number of fringes of decreasing magnitude are seen at low magnification (Fig. 3-8).

The first bright Fresnel fringe in a holey carbon film and how it is formed are shown in Figure 3-9. The appearance of the image of the edge of the hole changes in a through-focal series. At exact focus, no

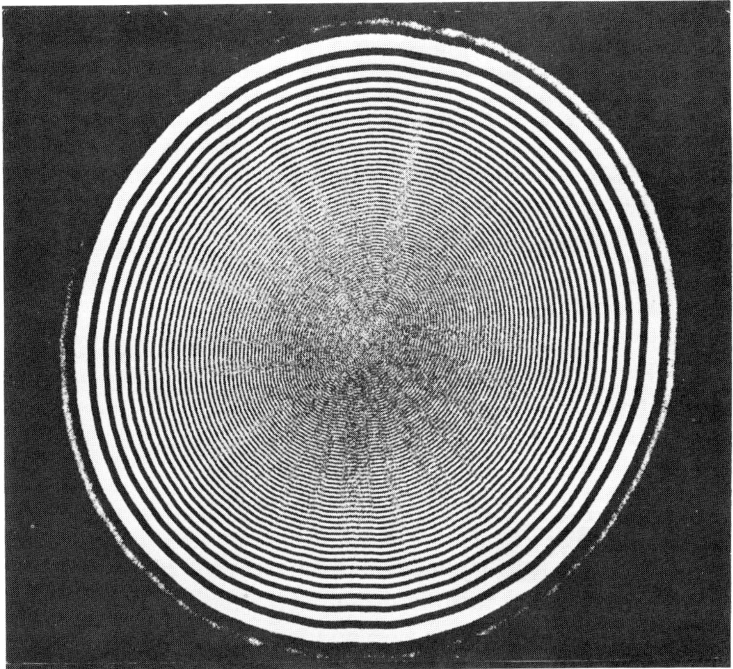

Figure 3-8. An underfocused image of a hole in a carbon film showing a train of Fresnel fringes inside the hole. A very small aperture, pointed filament, fully defocused condenser system, and a long exposure were used. Courtesy of Hitachi, Ltd.

fringe is visible (Fig. 3-9a). When the lens is overfocused (Fig. 3-9b) by increasing the objective lens current, the edge is seen as a bright band outlined with a dark ring. The magnitude of the fringe increases as the overfocusing is increased. If the current passing in the lens is reduced, the lens is underfocused and the edge is outlined by a bright fringe (Fig. 3-9c). This underfocused bright ring increases the contrast considerably and, with experience, a slight underfocusing can be used to enhance contrast of electron images, particularly at a low magnification. However, a loss of resolution occurs at this setting because the resolution is optimal only at the in-focus condition.

The Fresnel fringes are very useful in determining the presence of astigmatism and its correction in an EM. The fringes are also

peripheral (nonaxial) electrons are more strongly deflected and come to a focus ahead of those that are close to the axis (axial electrons) (Fig. 3-10b). The limit of resolution of a TEM by spherical aberration is given by the following equation:

$$d_s = k_s \cdot f \cdot \alpha^3$$

where d_s is the resolution due to spherical aberration; k_s is a dimensionless constant; f is the focal length; and α is the objective aperture angle in radians.

Glass lenses are corrected for spherical aberration by combining a divergent lens of different refractive indices with a converging lens. But the field strength across an electron lens cannot be equalized, and as such, this is the most important defect that limits the resolution of the TEM. The effect can be minimized by making the k_s, f, and α as small as possible. Given the following values, the resolution of a TEM due to spherical aberration can be calculated as follows:

$$k_s = 0.7; f = 3 \text{ mm}; \alpha = 4 \times 10^{-3} \text{ rad.}$$

so

$$d_s = 0.7 \times 3 \times 10^7 \times (4 \times 10^{-3})^3 \text{ Å} = 1.34 \text{ Å}.$$

Although the spherical aberration of a TEM is greatly reduced by making the aperture angle as small as possible, α cannot be reduced drastically. This would reduce the resolution due to increased diffraction effect. In Abbé's equation, the α should be as large as possible for a minimal diffraction effect, but it is just opposite in the spherical aberration. Therefore, the choice of α must represent a compromise between diffraction and spherical aberration effects. A very small α would also reduce the information-gathering power of the lens. The optimal α in a TEM (taking into consideration the diffraction and spherical aberration effects) is approximately 4.5 to 5×10^{-3} radians. Since the spherical aberration in the LM can virtually be eliminated, such a compromise is not necessary and the largest feasible α can be used.

Chromatic Aberration

When light rays of different wavelengths (hence colors) pass through a glass lens, a light of short wavelength is deflected more strongly than is the light of a long wavelength. Thus, blue light (short wavelength) is focused closer to the lens than red light of long wavelength (Fig. 3-11a). This defect, however, can be largely corrected by employing monochromatic light rays and by using glasses of varying refractive indices. When an objective glass lens is largely corrected for chromatic aberration, it is called *achromatic*; if it is corrected for both chromatic and spherical aberrations, it is called *apochromatic*.

The wavelength of an electron varies with the accelerating potential and the $\lambda = \frac{12.3}{\sqrt{V}}$ Å. Therefore, any change in the high voltage accelerating the electrons also changes the velocity of electrons. Chromatic effect, similar to the LM, occurs in the TEM when electrons of differing velocities (hence different wavelengths) are focused at different points in the image plane (Fig. 3-11b). However, unlike light rays, electrons of higher velocity (shorter wavelengths) are deflected less by the electron lens and are focused away from the lens. The slow-moving electrons (less velocity) of longer wavelengths, on the other hand, are acted upon by the lens field for a longer time and are focused closer to the lens. The high velocity electrons form a larger image than the low velocity electrons. This magnification change is known as *chromatic change in magnification*. The effect is conspicuous when a thick specimen is examined at a low magnification with a lower accelerating potential. The image is usually in focus at the center but is out of focus at the periphery.

The reasons for chromatic aberration in the TEM are (1) fluctuation in accelerating potential occurs due to variation in high

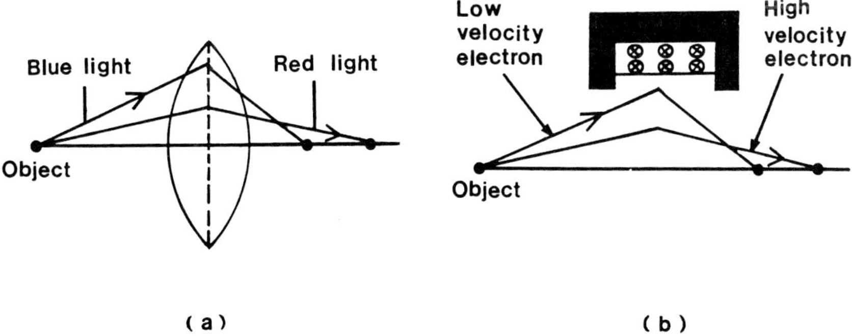

Figure 3-11. Chromatic aberration (a) in glass lens and (b) in electron lens.

voltage supply. This is usually controlled by an electronic high voltage stabilizer. Dirt in the electron gun causing discharge may also vary the accelerating voltage; (2) electrons emitted from the filament may have different initial velocity due to random emission. However, this is negligible at the range of 50 to 100 KV; (3) some velocity loss occurs due to inelastic scattering when the electrons strike the specimen. Although this mechanism causes chromatic aberration, it is a desirable effect for image contrast (*see* Chap. 4). Therefore, chromatic aberration is, to a certain extent, unavoidable in electron microscopy. Excessive voltage change with thick specimens, however, results in a poor resolution due to increased chromatic aberration; and (4) high voltage or lens current fluctuation will cause a change in the focal length of the objective lens and result in chromatic aberration. The equations for the resolution due to chromatic aberration are as follows:

$$(1) \quad d_{cv} = k_c \cdot f \cdot \alpha \cdot \frac{\Delta V}{V}$$

$$(2) \quad d_{ci} = 2k_c \cdot f \cdot \alpha \cdot \frac{\Delta I}{I}$$

where d_c and d_{ci} = chromatic effects considering voltage and lens current respectively; k_c = dimensionless constant; f = objective focal length; α = aperture angle; ΔV = change in voltage related to accelerating potential V; ΔI = change in objective lens current related to lens current I; and $\frac{\Delta V}{V}$, the ratio of high voltage change = $\frac{d_{cv}}{k_c \cdot f \cdot \alpha}$, and $\frac{\Delta I}{I}$, the rate of change in lens current = $\frac{d_{ci}}{2k_c \cdot f \cdot \alpha}$.

Therefore, it is necessary to stabilize the high voltage and objective lens current to a few parts per million for achieving a high resolution (3 to 5Å).

Astigmatism

This aberration is caused by an uneven magnetic field, particularly due to an asymmetry in polepieces. Therefore, the focal length of the lens differs in two directions at right angles. As a result, the lens focuses the electron beam preferentially in one direction, and an object point is not imaged as a point. Instead, it is imaged as two mutually perpendicular lines (Fig. 3-12). The defect is due to

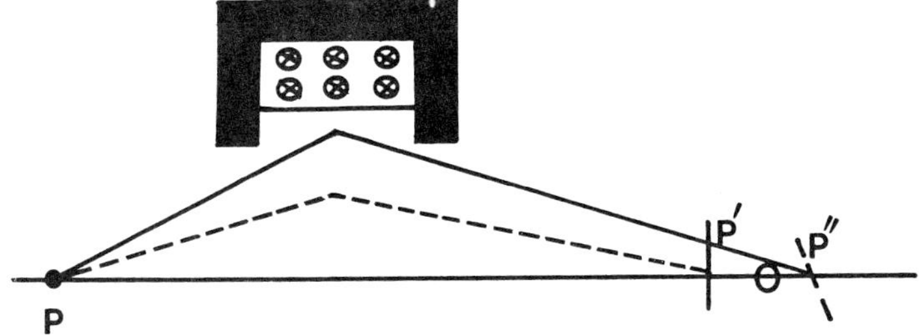

Figure 3-12. Image formation in an astigmatic lens. Point object P is imaged as two lines P′ and P″. The smallest circular image is between the two focal lines.

imperfect machining of polepieces of short focal lengths with small circular bores and inhomogenities in the iron. It can also be caused by contamination deposits in the polepiece and in the objective aperture. The defect can effectively be corrected by adding a compensating cylindrical lens called a *stigmator*.

Astigmatism is one of the most important defects that limits the resolution of an electron microscope, and the objective lens must be equipped with a stigmator. Such a lens is called a *compensated objective*. Stigmators are also provided in the second condenser lenses in the illuminating system of modern TEMs. A stigmator may be mechanical magnetic, electromagnetic, or electrostatic. The procedure for compensating astigmatism of the objective with a holey film is described in Chapter 7.

Distortion

Distortion is caused by the spherical aberration in an intermediate or projector lens that results in a variation of magnification throughout the image plane, but this does not affect the resolution. The outer part of an electron lens is much stronger than the central portion. Therefore, peripheral rays are deflected more strongly than the central rays. This gives rise to an increase or decrease in magnification of the image with increasing distance from the axis. The effect is striking when a square grid is examined at a low magnification. There are three types of distortion: (a) pincushion distortion—The aberration in this case results in an increase in magnification as the distance from the axis increases (Fig. 3-13a). It

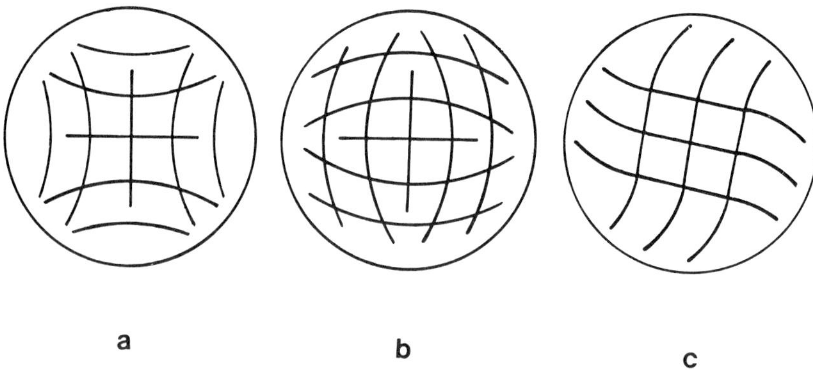

Figure 3-13. Distortion at a low magnification. (a) Pincushion distortion. (b) Barrel distortion. (c) Spiral distortion.

is noticed when the intermediate projector is working at a low strength while magnifying; (b) barrel distortion—This distortion is the opposite of pincushion distortion, higher magnification at the axis, lower at the periphery (Fig. 3-13b). This is due to the demagnification of the image caused by the intermediate lens while working at low strength; and (c) spiral or anisotropic distortion. The rotation of an electron in a magnetic field (*see* Chap. 2) can vary depending on the distance of a point from the axis that results in an image spiral (Fig. 3-13c).

The effect of distortion is negligible at high magnification. At this magnification, imaging is restricted to a very small central area, and off-axis rays are excluded. The intensity of illumination varies due to distortion and causes uneven exposure of photographic plates at low magnification. Manufacturers usually vary the strength of imaging lenses to achieve minimal distortion.

Space Charge Distortion

If the electron beam is highly concentrated, the electrons striking a specimen will repel each other (like charges repel) and cause the beam to spread out. Apertures are used to reduce the number of electrons, and a double condenser system is used for allowing a weak beam to illuminate the object. The effect of space charge distortion is noticed when the condenser is taken through crossover (*see* Chap. 5) that results in a change of focus in the image. It is, therefore, a bad practice to operate the TEM with the second condenser set at crossover.

Aberrations such as coma and curvature of the field are insignificant in electron microscopy. The resolution of a TEM is thus limited by diffraction, spherical aberration, chromatic aberration, and astigmatism. The methods for determining the resolution of the TEM are described in Chapter 8.

SELECTED BIBLIOGRAPHY

Agar, A.W., Alderson, R.H., and Chescoe, D.: *Principles and Practice of Electron Microscope Operation.* In Glauert, A. (Ed.): *Practical Methods in Electron Microscopy*, vol. 2. Amsterdam, North-Holland, 1974.

Chapman, S.K.: *Understanding and Optimising Electron Microscope Performance.* A Perkin-Elmer EM Publication.

Fischer, R.B.: *Applied Electron Microscopy.* Bloomington, Indiana University Press, 1954.

Hall, C.E.: *Introduction to Electron Microscopy*, 2nd ed. New York, McGraw-Hill, 1966.

Hardy, A.C. and Perrin, F.H.: *Principles of Optics.* New York, McGraw-Hill, 1932.

Meek, G.A.: *Practical Electron Microscopy for Biologists*, 2nd ed. London, Wiley, 1976.

Sjöstrand, F.S.: *Electron Microscopy of Cells and Tissues*, vol. 1. *Instrumentation and Techniques.* New York, Academic Press, 1967.

Wischnitzer, S.: *Introduction to Electron Microscopy*, 3rd ed. New York, Pergamon, 1981.

Zeitler, E.: Resolution in electron microscopy. *Adv Electronics Electron Phys, 25*: 277, 1968.

Chapter 4

DEPTH OF FIELD AND FOCUS, CONTRAST, AND IMAGE FORMATION

A HIGH-PERFORMANCE MICROSCOPE should have not only a high resolution but also a large depth of field, and the specimen should produce enough contrast for studying the image characteristics.

DEPTH OF FIELD

Although a lens may be set to give the sharpest image for a certain distance in the object plane, points located slightly in front or behind this plane will also be in fairly good focus. The range of distance over which an object can appear reasonably sharp is called *depth of field.* Qualitatively, it is the thickness of a specimen that may be in focus at any one time.

We know that a point object, when imaged through a lens, appears as a disk due to diffraction and lens aberrations (Chap. 3), and the diameter of this disk is the resolution of the lens. The depth of field is illustrated in Figure 4-1. The lens in this case is set to provide the sharpest image, P′, of a point object P (plane of exact focus). Points Q and R are equidistant from P. From geometrical grounds, the point Q then would appear as an out of focus disk in the image plane. This would also be the case with the point R. However, P′, the image of P, would also be a disk, the diameter of which is equal to the resolution limit of the lens. Therefore, as long as the disks of Q and R are equal to or smaller than that of the disk of P, which is in exact focus, all these points will be imaged under

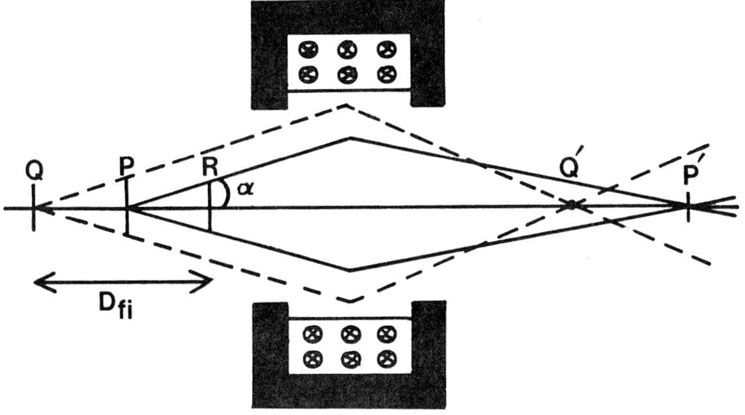

Figure 4-1. Depth of field (D_{fi}).

conditions of optimal resolution. The distance, D_{fi}, defined as depth of field is

$$D_{fi} = \frac{d}{\tan \alpha} = \frac{d}{\alpha}$$

where d = resolution limit; α = objective aperture angle. The equation also shows that the smaller the α, the larger the D_{fi}. We can then calculate the depth of field for light and electron microscopes.

For the light microscope (LM):

$$d = 2{,}000\,\text{Å};\ \tan \alpha\ (\text{for } 70°) = 2.75.$$

Therefore

$$D_{fi} = \frac{2{,}000}{2.75} = 727\,\text{Å}.$$

For a transmission electron microscope (TEM):

$$d = 5\,\text{Å};\ \text{and } \alpha = 4 \times 10^{-3}\ \text{rad}.$$

$$D_{fi} = \frac{5}{4 \times 10^{-3}}\ \text{Å} = 1{,}250\,\text{Å}$$

The depth of field of the LM is considerably less, indeed less than its resolution. On the other hand, the depth of field of a TEM is very large.

The large depth of field in the TEM has certain advantages over the LM. The maximum thickness of specimen that the electrons can penetrate is generally less than the depth of field of the microscope. Thus, if any part of the specimen is in focus, the entire specimen is usually in focus. A large D_{fi} is also good for stereoscopy, and specimen positioning and focusing are not as critical as in the LM. The large D_{fi} can be advantageous for transmission electron microscopic examination of cellulose fibers and other similar materials, which are difficult to position in one plane.

The advantage of a large depth of field in a TEM over the LM, however, cannot be overemphasized. A large D_{fi} of the TEM gives rise to superimposition effect in a thick specimen. An LM with a small D_{fi} can be focused at several depths in such a specimen. Different cross sections of the objects can be examined by using lenses of different focal length (low, high dry, and oil immersion), and information corresponding to structures in the third dimension can be obtained. This is not possible with the TEM.

DEPTH OF FOCUS

An equivalent distance corresponding to the depth of field on the object side is also found on the image side. *Depth of focus* is the range of distance over which the image is equally sharp. This is shown in Figure 4-2 and can be calculated by the following equation:

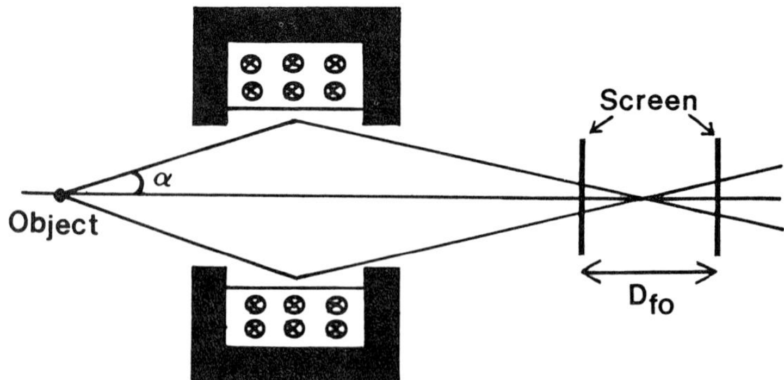

Figure 4-2. Depth of focus (D_{fo}).

$$D_{fo} = \frac{d \cdot M^2}{\alpha}$$

where D_{fo} = depth of focus; d = resolution limit; M = total magnification (product of the magnification of all imaging lenses); α = objective aperture angle. We can then calculate the D_{fo} for a TEM from the following data:

$$d = 5\text{Å};\ \alpha = 5 \times 10^{-3}\ \text{rad.};\ M = 100{,}000\times$$

so

$$D_{fo} = \frac{5 \times (100{,}000)^2}{5 \times 10^{-3}}\ \text{Å} = 1{,}000\ \text{m}$$

= 1 km, which for all practical purposes is infinite.

A large depth of focus of the TEM makes the job of focusing the image simple, especially at high magnification. Also due to this enormous value, the fluorescent screen, the photographic plate, or a 35 mm film can be placed anywhere below the projector lens. A 35 mm camera is placed immediately below the projector lens in some electron microscopes, e.g. Philips. Although the magnification on this film will differ, the image will be as sharp as that on the photographic plate placed below the screen.

CONTRAST AND IMAGE FORMATION

Contrast is the relative difference in intensity between the object and its immediate surroundings.

$$\text{Simply, contrast} = \frac{\text{Intensity of background} - \text{intensity of object}}{\text{Intensity of background}}$$

The eye is sensitive to amplitude contrast (intensity difference), and it can distinguish an intensity difference of 10% or more. It is also sensitive to wavelengths of different colors (color contrast), and it can interpret LM images directly. For transmission electron microscopy, the specimen must be thin enough to allow enough electrons to pass through so that they can carry the information about the specimen. This information cannot be interpreted directly by the eye. It can be converted only to amplitude and phase contrasts, but the eye is insensitive to phase contrast. The amplitude contrast is formed on a fluorescent screen or on a photographic plate. Direct visualization of electron images on the screen generally provides less information than a permanent photographic recording. Unlike light, different wavelengths of electrons cannot be translated into different colors. Thus, the electron micrographs are usually black and white.

Contrast in Light Microscopy

Four processes, viz., absorption, interference, diffraction, and scattering, are involved in the image-forming process in light and TEMs. Although all these mechanisms together generally contribute to contrast and image formation, absorption is the most predominant process in light microscopy. Differential absorption of light throughout the specimen causes a variation in intensity and/or color of the image. Different stains are often used to accentuate this mechanism in light microscopy. Interference produces phase effects, but the eye is not sensitive to it. Although the diffraction effect can be used to enhance contrast in light and electron images, it also causes a loss of resolution.

Contrast in Electron Microscopy

Scattering, which plays very little role in light microscopy, is the most important process for contrast and image formation in the TEM.

Interaction of Electron Beam with Specimen and Scattering

Incident beam electrons interact with the specimen, which gives rise to many reactions, some of which are beneficial, while others are harmful. These reactions are as follows:

TRANSMITTED UNSCATTERED ELECTRONS: These electrons pass through the specimen without being deflected or without losing energy (Fig. 4-3a) and, therefore, form *pure beam electrons*.

ELASTIC SCATTERING: When the incident electron strikes the nucleus of a specimen atom or passes very close to it, the incident electron is deflected through a large angle without loss of energy (Fig. 4-3b). This is because the atomic nucleus is much heavier than the electron. This process is called *elastic scattering*. The amount of

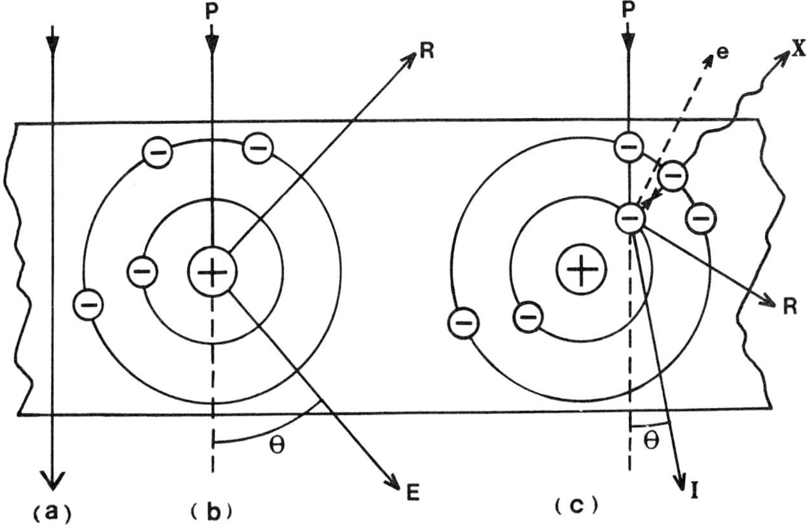

Figure 4-3. Interaction of incident electron beam with the specimen and electron scattering. (a) Unscattered transmitted electrons. (b) Elastic scattering. (c) Inelastic scattering. P = incident primary electron; E = elastically scattered electron; I = inelastically scattered electron; X = characteristic x-rays; e = secondary electron; and R = reflected or backscattered electron.

this scattering increases in a thick specimen as the probability of striking a nucleus is increased; it is also proportional to the atomic number of the nucleus.

INELASTIC SCATTERING: If the incident electron collides with an electron orbiting around an atomic nucleus of the specimen, the incident electron is deflected through a small angle with a loss of velocity (Fig. 4-3c). This is called *inelastic scattering.* This is an encounter of two particles of the same mass. Therefore, they will share their velocities (due to laws of conservation of momentum). The energy imparted to the orbital electron (hence the atom of the specimen) by the incident electron raises that atom to an excited energy state, which usually causes radiation damage to the specimen (*see* Chap. 18). Inelastic scattering is much more frequent than elastic scattering because there are far more electrons than nuclei in a specimen. A reduced velocity of incident electron results in a change in its wavelength and causes chromatic aberration. The thicker the specimen, the more the energy loss and poorer the resolution.

The amount of scatter is directly related to the thickness of the specimen; the incident electron encounters a large number of atoms in thick samples. More electrons are present around a large and heavy atomic nucleus. The size of the atomic nucleus is proportional to the atomic number and, roughly, to the physical density. The higher the density, the greater the probability of scatter. The total scatter in a specimen is directly proportional to its *mass density*, which is the product of the thickness and density. It is expressed as $\mu g/cm^2$. Although a thick specimen scatters more electrons than a thin specimen, it is the physical density, not the thickness, that gives more contrast. A 7Å thick platinum deposit, e.g. (atomic number $(Z) = 78$, density = 21.45) gives approximately the same amount of contrast as a chromium deposit of 18Å ($Z = 24$, density = 6.92).

The mass density of a carbon film, 50Å thick (density = $2\ g/cm^2$) $= 50 \times 10^{-8} \times 2 = 1 \times 10^{-6}\ g/cm^2 = 1\ \mu g/cm^2$.

Electron beam-specimen interaction also gives rise to other reactions (Fig. 4-3). These are as follows:

SECONDARY ELECTRONS: These electrons are emitted from the specimen due to energy imparted during a collision between an incident electron and an orbital electron. They are employed to form an image in the scanning electron microscope (SEM).

CHARACTERISTIC X-RAYS: These are produced when a secondary electron is emitted. An electron from another orbit replaces the lost electron (orbital jump) and gives off a quantity of energy in the form of characteristic x-rays. They can be used for x-ray microanalysis.

BACK SCATTERED OR REFLECTED ELECTRONS: These electrons of high energy are produced when an elastic or inelastic scattering occurs at the specimen surface. Instead of passing through the specimen, the primary electrons are emitted from its surface. They can be used for imaging in the SEM.

IONIZATION AND CHARGING: Ionization occurs when an electron of the specimen atom is dislodged by an incident electron (*see* Chap. 18). When electrons are transmitted through a thin specimen, it can accumulate charge by loss of secondary electrons. This can destroy a nonconducting specimen.

HEATING: Transfer of heat to the specimen occurs during inelastic scattering, and the specimen becomes hot. This causes thermal drift and movement of the specimen and results in blurring of the image. The temperature of nonconducting materials may reach 100 to 200° C in the TEM.

Contrast Due to Absorption

When the incident electron loses all its energy to the specimen by the inelastic scattering process, it is absorbed by the specimen. Absorption produces amplitude contrast, but this is negligible with a thin specimen. Absorption in a thick specimen, however, can cause severe thermal drift.

Objective Aperture and Contrast in Transmission Electron Microscopy

Electrons are scattered when a narrow pencil of electrons strikes a specimen. Therefore, the narrow pencil is widened after the electrons are transmitted through the specimen. The electrons then encounter the objective aperture, a very small circular hole (10 to 80 μm) in a thin disk or foil of nonmagnetic metal such as platinum or molybdenum.

The aperture defines the acceptance angle of the lens (numerical aperture) and allows the scattered electrons to travel through. However, the electrons that are scattered by an angle larger than that defined by the objective aperture are taken out of the beam and fail to reach the corresponding image point. The loss of these electrons would cause a loss of intensity at these points, and the areas would appear dark in the image. The less dense (more electron transparent) areas of specimen would scatter fewer electrons, and thus appear brighter. This results in a variation of intensity in the image of the specimen. The objective aperture, therefore, due to its subtractive action, is the principal source of contrast in the TEM (Fig. 4-4). Its importance can be appreciated by suddenly withdrawing it while examining a well-focused image. The image almost disappears due to a drastic reduction in contrast.

The electrons scattered by an angle larger than the aperture angle, α, strike the metal surface of the objective aperture and are taken out of the beam by being conducted back to the earth. Otherwise they will reduce the contrast and degrade the image. Ideally, this aperture should be placed at the center of the lens. However, due to the

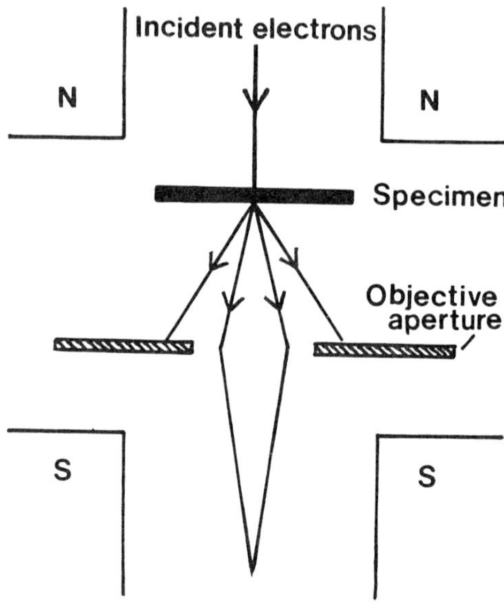

Figure 4-4. Action of the objective aperture in image formation by the transmission electron microscope.

position of the specimen holder, the aperture is usually placed close to the back focal plane.

The smaller the aperture (hence a smaller α), the greater the number of electrons lost and the better the contrast. However, small apertures (less than 20 μm) become contaminated quickly, and the contamination deposits become electrically charged by electron bombardment. Thus, they would act as small electrostatic lenses, introduce asymmetry in lens field, and cause astigmatism. Small apertures are also very difficult to clean and difficult to center. Therefore, electron microscopists prefer an aperture of 40 to 50 μm dia. An aperture larger than the size of 80 μm contributes little contrast, if any. Calculation of the aperture angle follows:

$$\alpha_o = \frac{\text{Diameter of objective aperture}}{2 \times f} \text{ rad.}$$

where α_o = objective aperture angle; f = focal length. Therefore, for a TEM with an objective aperture of 40 μm dia, f of 4 mm:

$$\alpha_o = \frac{40}{2 \times 4 \times 10^3} \text{ rad.} = 5 \times 10^{-3} \text{ rad.}$$

Elastically scattered electrons may or may not appear in the final image. They are scattered through a very large angle, and the spherical aberration of the objective lens affects them more than the inelastically scattered electrons. Thus, they are generally removed from the beam. Although a large object can produce contrast by elastic scattering, contrast of small objects in the TEM is mainly due to inelastic scattering.

Defocus Contrast

This contrast is observed in an out-of-focus image and is due to the formation of Fresnel fringes (*see* Chap. 3). The contrast of an in-focus image is drastically reduced when the objective aperture is withdrawn. However, if the objective current is now varied on either side of the in-focus point, the contrast on the underfocus and overfocus side is greatly increased. The contrast is minimal in the in-focus condition and maximal on the underfocus side (Fig. 4-5). Therefore, many electron microscopists take advantage of this enhanced contrast by recording the image at a slightly underfocus condition. This, of course, causes a slight loss of resolution. The

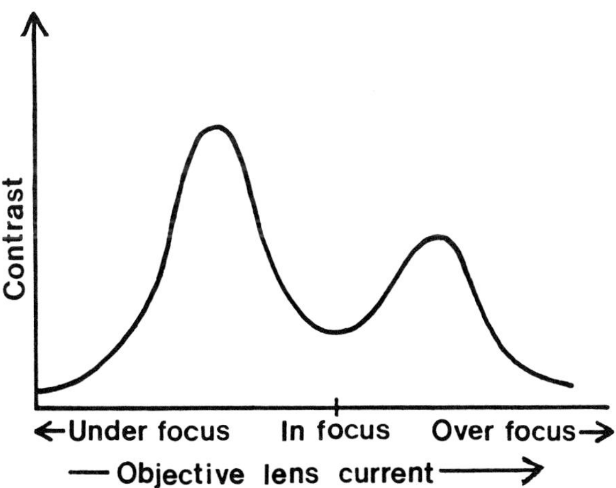

Figure 4-5. Image contrast and objective lens focus.

amount of underfocusing to be used is gained by experience in transmission electron microscopy.

Objective Focal Length and Contrast

Some TEMs use top-entry specimen holders, and the height of these holders can be adjusted to obtain an increase in the focal length of the objective lens. Thus, the image contrast is increased without resorting to a smaller aperture. The effective aperture angle (α_o) is decreased due to the increased focal length (*see* equation for α_o in this chapter), which results in a better contrast. However, this will cause a loss of resolution and the magnification will be lower. The relationship of contrast to the objective focal length is illustrated in Figure 4-6. The focal length of the objective lens can be varied from 2 mm to 14 mm, and the contrast will be almost 12 times higher at 14 mm than at 2 mm. However, the resolution and magnification will be approximately 10 times lower with an objective of this focal length.

Accelerating Voltage and Contrast

The higher the accelerating KV, the lower the scattering angle of electrons and vice versa. Therefore, with a given size of aperture, a reduction in accelerating potential will result in a greater number of

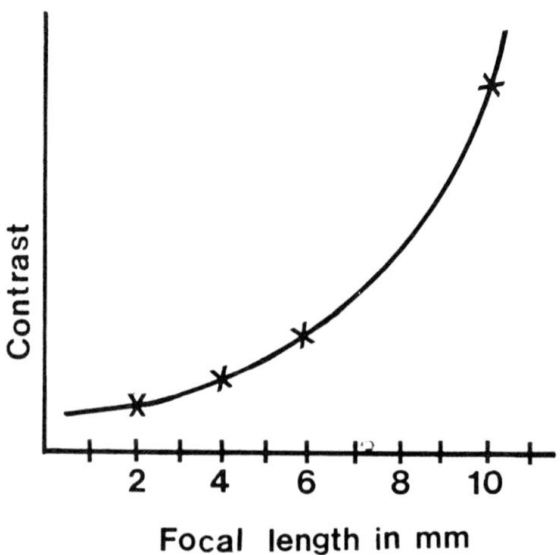

Figure 4-6. Effect of increased objective focal length on contrast.

electrons lost, and a better contrast. More electrons pass through the objective aperture at a higher KV, resulting in a specimen glare and loss of contrast. However, the specimen has to be very thin for the low potential electrons to penetrate it, and the resolution is less at lower KV than at higher KV.

Phase Contrast

This contrast is apparently produced by an interference effect of scattered electrons due to the aperture angle and the spherical aberration of the objective lens. Elastically scattered electrons pass through the aperture and combine with unscattered transmitted electrons in the image plane, which produces an extremely fine granularity in the image. This phase effect is detected only at the highest magnification, very close to the focus and close to the highest resolution.

Electron Noise

Electron noise is the statistical variation in arrival of electrons at the photographic plate, which produces a random contrast independent of the specimen. This is caused by an inhomogeneity of the electron beam due to random emission of electrons from the filament, and the random contrast can be photographed in the absence of a specimen. The granularity caused by electron noise can be reduced by collecting more electrons on a photographic plate at a higher magnification (more than 30,000×).

Photographic Density and Contrast

The contrast in electron micrographs can be altered by increasing or decreasing the exposure time (hence by collecting more or fewer electrons) on a photographic plate. A moderately dense negative provides a good contrast (*see* Chap. 9).

Contrast in Biologic Specimens

These specimens generally have a low physical density, and their electron scattering power is very low. Therefore, they have to be shadowed with heavy metals, stained positively or negatively so that they can scatter more electrons (*see* Chap. 11). Dark-field operation also enhances the image contrast (*see* Chap. 16).

A biologist faces several problems in achieving the highest resolution and maximum contrast he wants in a TEM. A thin specimen would give a better resolution but poorer contrast, and in a thick specimen, vice versa. Similarly, a higher accelerating potential will provide a better resolution but a poorer contrast. The effect is just the opposite with a low accelerating potential. Since contrast in biologic samples is more important than the resolution, a compromise has to be made. Therefore, the full resolving power of the modern high-performance TEMs can rarely be used for biologic specimens. Often, the biologist has to be satisfied with a routine resolution of 10Å and a good contrast.

SELECTED BIBLIOGRAPHY

Agar, A.W., Alderson, R.H., and Chescoe, D.: *Principles and Practice of Electron Microscope Operation.* In Glauert, A. (Ed.): *Practical Methods in Electron Microscopy*, vol. 2. Amsterdam, North Holland, 1974.

Chapman, S.K.: *Understanding and Optimizing Electron Microscope Performance.* A Perkin-Elmer EM Publication.

Fischer, R.B.: *Applied Electron Microscopy.* Bloomington, Indiana University Press, 1954.

Hall, C.E.: *Introduction to Electron Microscopy*, 2nd ed. New York, McGraw-Hill, 1966.

Lovell, D.J. and Chapman, S.K.: The effect of variations in objective focal length on electron microscope performance. *J Micros 105*:277, 1975.

Meek. G.A.: *Practical Electron Microscopy for Biologists*, 2nd ed. London, Wiley, 1976.

Sjöstrand, F.S.: *Electron Microscopy of Cells and Tissues*, Vol. 1, *Instrumentation and Techniques.* New York, Academic Press, 1967.

Wischnitzer, S.: *Introduction to Electron Microscopy*, 3rd ed. New York, Pergamon, 1981.

Chapter 5

THE TRANSMISSION ELECTRON MICROSCOPE AND ITS PARTS

AMONG THE DIFFERENT TYPES of electron microscopes, the one most widely used is the transmission electron microscope (TEM). As the name implies, electrons transmitted through a thin specimen form the image in the TEM. The lenses of TEMs are generally of the magnetic type, and the design of the modern electromagnetic TEM is described here. The column of this microscope basically consists of three parts: (1) the illuminating system; (2) the image-forming system; and (3) the image-translating system. The column is usually mounted vertically for providing mechanical rigidity. The illuminating system is generally at the top, and the imaging and image-translating systems are at the bottom. A modern high-resolution TEM with attachments for scanning and scanning transmission electron microscopy and the cross section of the total system are shown in Figure 5-1.

THE ILLUMINATING SYSTEM

This contains the electron gun, which is the source for electrons, and a condenser system that directs the electron beam onto the specimen and regulates the intensity.

The Electron Gun

The electron gun consists of the filament, shield, and anode. It acts like a positive electrostatic lens. The filament and shield can be connected to the negative terminal of high voltage (HV) supply by

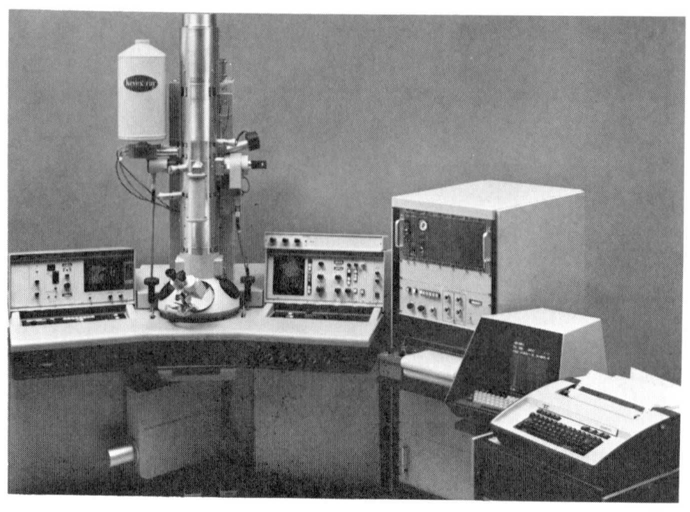

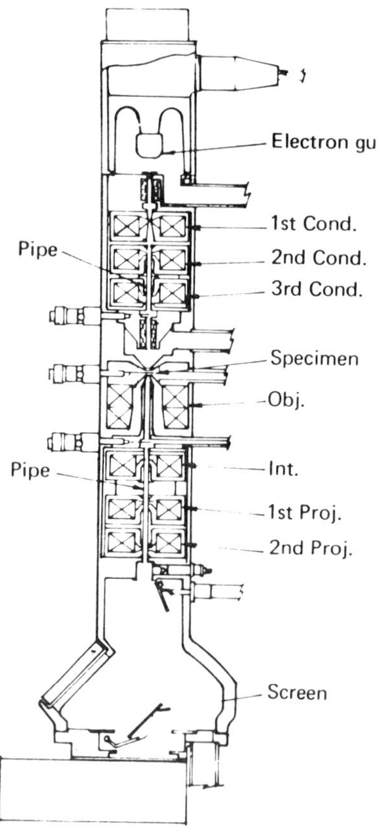

Figure 5-1. (a) A modern high-performance transmission electron microscope (Hitachi H-600) with attachments for scanning, scanning transmission electron microscopy, and X-ray microanalysis—a total analytic system transmission electron microscope. (b) A cross section of the microscope column. Courtesy of Hitachi, Ltd.

two ways. In the nonbiased gun, the filament and shield are connected directly to the HV through a pair of balancing resistors. In the self-biased system, the filament is connected to the HV supply through a bias resistor and a pair of balancing resistors, but the shield is connected to the HV directly. The filament life in a nonbiased system is very short, and this system is no longer employed in electron microscopes. Thus, it will not be discussed here. The gun top is hinged for rapidly changing a burned-out filament, and the cathode assembly plugs in the high voltage insulator (Fig. 5-2).

Filament (Cathode, Thermoionic Emitter)

A *V*-shaped pure tungsten wire (Fig. 5-3a) approximately 0.1 mm in diameter emits the electrons. It is connected to the negative terminal of HV supply and is less negative than the shield. The required work function for emitting the electrons from the metal surface is provided by increasing the temperature of the filament (thermoionic emission, *see* Chap. 1). When an electric current

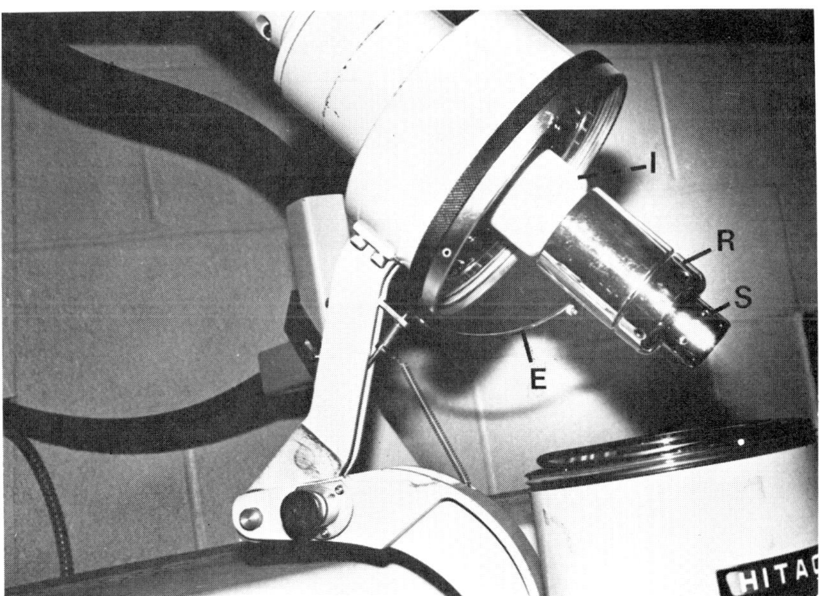

Figure 5-2. An electron gun top opened for filament changing. S = shield; R = set ring; I = high voltage insulator; and EB = earthing ring.

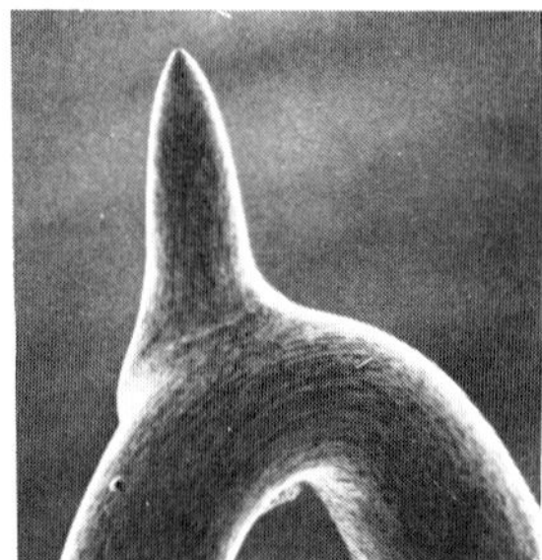

Figure 5-3. Tungsten filaments. (a) A standard hairpin filament. (b) A pointed filament.

(filament current) is applied to the filament, its tip becomes red hot, from which a copious supply of electrons can be obtained. These electrons carry electric charges to the surface of the anode, and an electric current known as *beam current* flows between the cathode and anode.

The electron source must be small because the larger the source the greater the electric power consumption. Only a small source can be focused on a small area of the specimen, and a large beam current produced by a large source overloads the HV supply. A very narrow pencil of electrons passes through the lens apertures and polepiece bores, for which the intensity of the beam must be very high. A symmetrical source is convenient for alignment.

The filament must have a reasonably long life, and the work function (*see* Chap. 1) of the metallic emitter must be low. The average life of a filament (actual filament burning time) varies from twenty to fifty hours. The life of the filament depends on the operating temperature of the filament and oxidation of tungsten due to the presence of traces of oxygen. The operating temperature for tungsten is approximately 2,600° K; a higher temperature increases evaporation of the metal and drastically reduces the filament life. Although tungsten has a work function of 4.5, which is higher than other metals, it is most suitable for electron emission. Thoriated tungsten and other oxide-coated metals having a work function lower than tungsten have been used. Their source is usually brighter than that of tungsten, but they are extremely susceptible to traces of air.

A coherent and extremely bright electron source can be produced by using a pointed filament and a small condenser aperture. A pointed filament is made of a tungsten spigot welded into a hairpin filament and then chemically etched to form a fine point (Fig. 5-3b). They have been claimed to reduce the overall contamination, but their life is considerably shorter than that of the conventional hairpin type of filament.

Shield (Wehnelt Cylinder, Grid, Cathode Cap)

Electrons are emitted at random in all directions from the cathode. These electrons are shaped into a beam by the shield, which has a circular aperture in the middle (Fig. 5-4). The tip of the filament is placed close to shield aperture. Filaments for some instruments are supplied precentered and adjusted to a correct height. In others, the filament centering and height adjustment relative to the shield are done by adjusting screws. The shield is directly connected to the negative terminal of the HV supply.

Anode (Positive Electrode)

This is an apertured disk that is connected to the positive terminal of the HV supply and is grounded like the other parts of the microscope column. The emitted electrons are accelerated in the

Figure 5-4. Cathode assembly. (a) Filament housing. (b) Cathode shield.

space between the filament and anode by falling through a potential drop of 20 to 125 KV.

The Self-biased Electron Gun

This system is schematically illustrated in Figure 5-5. When the beam current flows through the bias resistor, a negative potential between the cathode and shield is established, and the filament becomes less negative (−100 to −500 V) than the shield. The equipotential surfaces (*see* Chap. 2) concentrated in the space between the shield and the anode would then bulge toward the cathode in the opening at the shield (Fig. 5-5). Since the tip of the filament is located close to the shield aperture, the action of a strong converging electrostatic lens is produced. Thus, the electrons emitted from the filament tip are funneled to converge toward a point in front of the shield (toward the anode), and the *electron cloud* or *space charge* surrounding the filament tip is focused. The density of electrons per unit area is the highest at this point. This so called *crossover* becomes the actual source (rather than the filament) of electrons from which they can be drawn away by the anode.

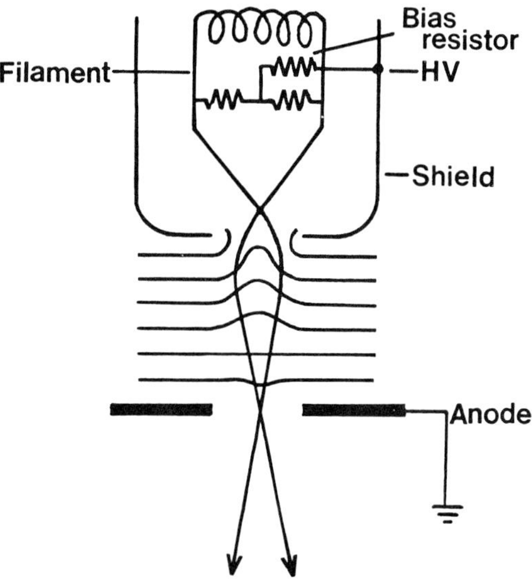

Figure 5-5. A biased electron gun. The filament is connected to the high voltage supply through a bias resistor; the shield is connected directly.

Crossover is defined as the condition when the image of the electron cloud is focused on the specimen. Since the filament tip is *V* shaped, the image of the crossover is elliptical, not circular.

Compared to the filament tip, the crossover represents a highly concentrated electron source from a small area. Therefore, an intense illumination with a low beam current is possible. This reduces the load on the HV supply.

Variable Self-biased Guns

These are now standard in modern TEMs. The bias in these guns can be varied through potentiometers. Thus, the intensity of illumination in these systems can be easily varied, and a low intensity beam can be used for specimens that are highly susceptible to beam damage. This is not possible with the self-biased system because its intensity is usually fixed. Its intensity, however, can be varied by adjusting the filament height in the cathode assembly.

Operation of a Self-Biased Gun

When a current is applied to heat the filament, the beam current rises as the filament current is increased (Fig. 5-6). The rise in beam current is rapid and steep at first; it then rises slowly, eventually reaching a flat maximum value at saturation (the beam current meter levels off). At this point, the beam current is independent of the filament current and the electron gun reaches the optimal

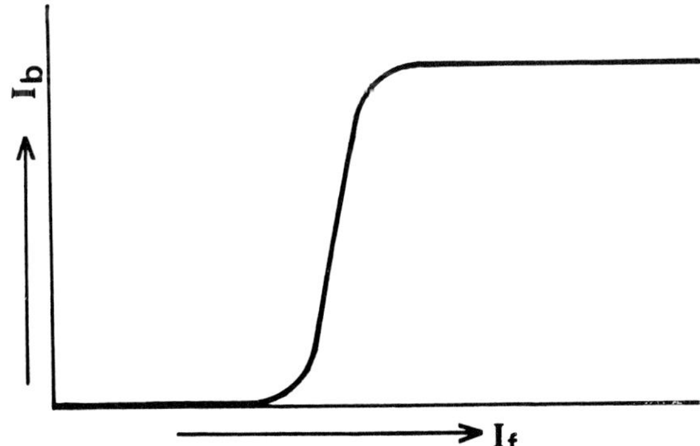

Figure 5-6. Operation of a self-biased gun. I_b = beam current; I_f = filament current.

operational level. Once the saturation of the gun is reached, which is quite rapid, further increase in filament current produces no change in the intensity; it only shortens the life of the filament.

As the current is applied to the filament, a faint spot appears on the screen at first. As the filament current is increased, an intense spot surrounded by two broken rings at the opposite ends is seen. A further increase in filament current makes the broken rings contract toward the central bright spot. Finally, a single solid bright spot is observed at saturation. The electron gun is a major source of x-rays, and it is shielded to protect the operator.

The Condenser System

The condenser lens focuses the illuminating beam on the specimen and controls the intensity of the beam. Its focal length is usually on the order of a few cm. The intensity is regulated by changing the focal length of the lens. A single condenser was provided in earlier instruments that produced an illuminating area similar to the size of the crossover (40 to 50 μm). This caused unnecessary heating of the specimen and, more important, contamination (*see* Chap. 8).

A *double condenser system* (Fig. 5-7) is now standard in modern TEMs. The aperture angle of illumination depends on the size of the condenser aperture, which is larger in the first condenser than in the second condenser. The first condenser lens produces a demagnified electron source, and the second condenser lens focuses it on the specimen. The first condenser is usually operated at a fixed strength, and the intensity is controlled by varying the focal length of the second condenser. The aperture angle, α_c, is limited by the physical aperture of the second condenser lens when the lens is defocused. The condenser apertures vary in size from 100 to 500 μm. The double condenser system also has polepieces. The second condenser lens is usually equipped with a stigmator to correct the astigmatism of this lens system. This astigmatism produces an uneven electron source and causes a loss of intensity. Its correction is described in Chapter 7.

The advantages of a double condenser lens system are that (1) a spot size as small as 2 μm can be used to expose the specimen to the beam at one time; therefore, the contamination is limited only to a small specimen area; and (2) it gives good illumination at higher magnification.

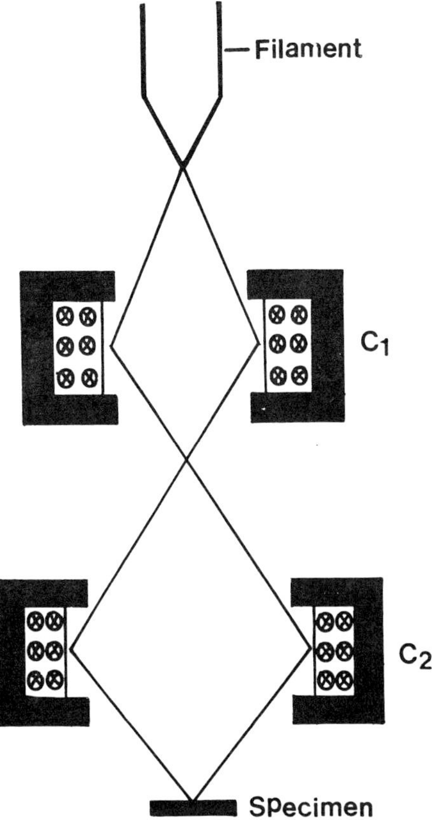

Figure 5-7. A double condenser system. C_1 = first condenser lens; and C_2 = second condenser.

THE IMAGE-FORMING SYSTEM

This system consists of three or four lenses that produce a magnified image of the specimen. The three-lens imaging system contains the objective and two other lenses for which different terminology is used by different manufacturers, e.g. intermediate and projector, or projectors 1 and 2. The four-lens imaging system is now almost universal in TEMs. It contains the objective and three other lenses, which again may be described by different terminology, e.g. intermediates 1 and 2, and projector, or intermediate, projectors 1 and 2. The addition of the fourth lens increases the flexibility and total magnification of the system.

The Objective Lens

This lens is the heart of the electron microscope, and it determines the ultimate performance of the instrument. It is responsible for the focus, resolution, contrast, and the primary magnification of the image. The requirements of this lens are that (1) it should have minimal aberrations and its focal length should be as short as possible for a high resolution; (2) it must have a physical aperture for contrast (*see* Chap. 4); (3) it must have a stigmator for correcting the astigmatism (Fig. 5-8) introduced by the objective polepiece; and (4) the specimen must be placed as close to the focal plane as possible. The objective aperture sizes of 20 to 50 μm are most commonly used, and molybdenum apertures are most suitable. Recently, self-cleaning thin foil gold apertures have been introduced.

Projector Lens(es)

The projector lens, as the name indicates, projects the final

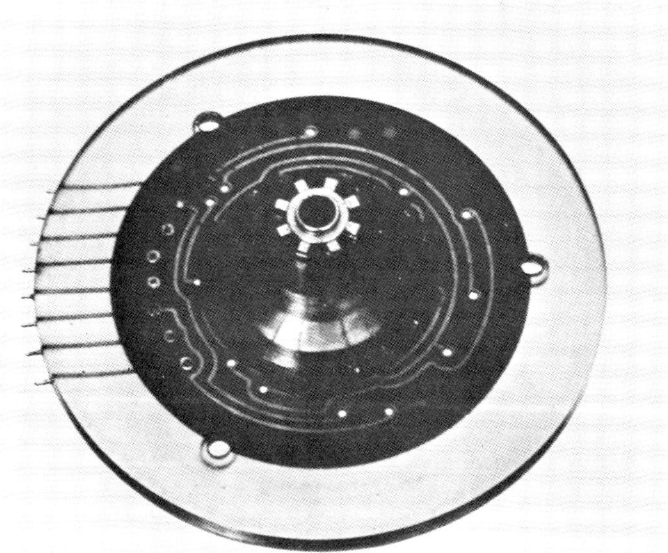

Figure 5-8. Objective lens octupole stigmator. From A.W. Agar, R.H. Anderson, and D. Chescoe, *Principles and Practice of Electron Microscope Operation*. In A. Glauert (Ed.), *Practical Methods in Electron Microscopy*, Vol. 2, 1974. Courtesy of Elsevier/North-Holland, Amsterdam.

magnified image on the screen or photographic film. One or two intermediate lenses are placed between the objective and projector for obtaining a very high magnification. Magnification can be changed by varying the strength of one of the intermediate lenses while the objective and projector lens strengths are fixed. The total magnification of the microscope is the product of the magnifications of the objective, one or two intermediates, and the projector lenses.

Although many modern electron microscopes boast direct magnification in the range of 500,000×, it is seldom if ever necessary to record the image at this magnification. The magnification required to resolve fine details of an image can be calculated from the maximum useful magnification (*see* Chap. 1). For a 2Å resolution, the overall magnification required is 10^6×. The photographic negative generally allows a 10 to 20× enlargement, and the maximum direct magnification needed on the screen is usually 250,000×. Distortion is caused by the spherical aberration present in the intermediate or projector lens (*see* Chap. 3).

IMAGE TRANSLATING SYSTEM

The electron image is translated into a visual image when the electrons bombard the fluorescent screen. Thus, the kinetic energy of the electrons is converted to light energy via fluorescence. A fluorescent screen is made of a thin metal plate coated with a thin layer of activated zinc or cadmium sulfide. It fluoresces greenish-yellow, and the eye is most sensitive to this color in the dark. The resolution on the screen depends on the grain size of the fluorescent material, which should be 100 μm or smaller. This resolution is usually 40 to 50 μm, and it is better than the resolution limit of the eye.

Therefore, a 5 to 10× magnifier (usually a binocular) can be used to examine the electron image in greater detail. The binocular allows the eye to approach the screen closer than the near point of the eye. The aperture angle of the binocular is also much higher than that of the eye. A very important advantage of the binocular is that while the optical magnification of the image is increased 5 to 10×, there is little loss of brightness to the image. Thus, the demand for the total magnification on the imaging system is reduced. If a direct electron optical magnification was used for obtaining this added magnification, the resultant loss in image brightness would be drastic. This is because the brightness of the image is inversely proportional to the square of the total magnification.

The graininess of the screen is much larger than that of the photographic emulsion. Therefore, the image should be photographed for achieving a better resolution (*see* Chap. 9). The fine grain silver bromide emulsion is sensitive to electron energy. The camera may be placed below the fluorescent screen or right after the projector lens (*see* Chap. 4). The screen has to be tilted for convenient viewing of the image through the binocular, which produces a distorted image. It is emphasized that the intensity for viewing the image should not be too high because it is harmful to the specimen and it increases the rate of contamination. At the same time, however, the intensity must be sufficient for studying the image and for photographic recording.

Alignment

A very narrow pencil of electrons (a few micrometers wide) passes through various lenses, small polepiece bores, and apertures. It is, therefore, necessary to put the axes of different lenses in line so that the axis of the illuminating beam coincides with that of the image-forming system. When the microscope has been properly aligned, the following optimal conditions are obtained at the center of the screen: (1) the illumination is bright and is centered as the intensity is varied; (2) the image does not sweep as the objective focus is varied; and (3) a change of magnification is obtained without the loss of the field or the illumination on the field. The procedure for alignment is described in Chapter 7.

OTHER COMPONENTS OF THE ELECTRON MICROSCOPE

In addition to the basic parts described in this chapter, the TEM also contains the following components.

Condenser and Objective Aperture Holders

The design of the movable aperture holders for the second condenser and objective lenses is identical (Fig. 5-9). Most modern electron microscopes use a molybdenum foil punched with three or

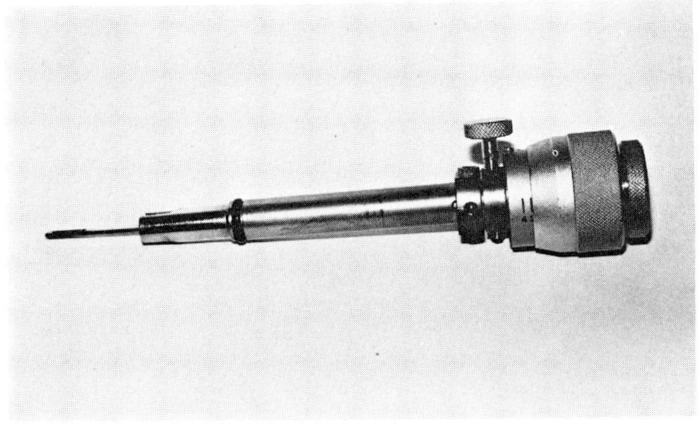

Figure 5-9. Objective aperture holder.

four holes for this purpose. Thin foil molybdenum apertures are more suitable than platinum apertures because they do not become contaminated as fast as the platinum apertures. The apertures can be centered by micrometer controls with a clickstop mechanism located outside the column. The molybdenum apertures are cleaned by heating them on a tungsten boat in a vacuum evaporator. They cannot be heated on a Bunsen flame. Platinum apertures can be cleaned like molybdenum apertures or they can be heated on a Bunsen flame and then immediately dipped into hydrofluoric acid. They are washed in double distilled water and a solvent (alcohol or petrol ether) before inserting into the TEM.

The Specimen Stage and Holders

To examine the whole area of the specimen, the movement of the specimen must be extremely precise, with a minimal backlash, especially at a higher magnification. The whole stage along with its controls must remain stationary during photographic recording. The stage should also make good thermal contact with the specimen holder to dissipate heat. Therefore, the requirements for the construction of these stages are very stringent. The stage is made of nonmagnetic materials and is usually mounted in or above the objective lens, and it can be moved in *X* and *Y* axes (Fig. 5-10).

The specimen holders can be of either top-entry or side-entry type

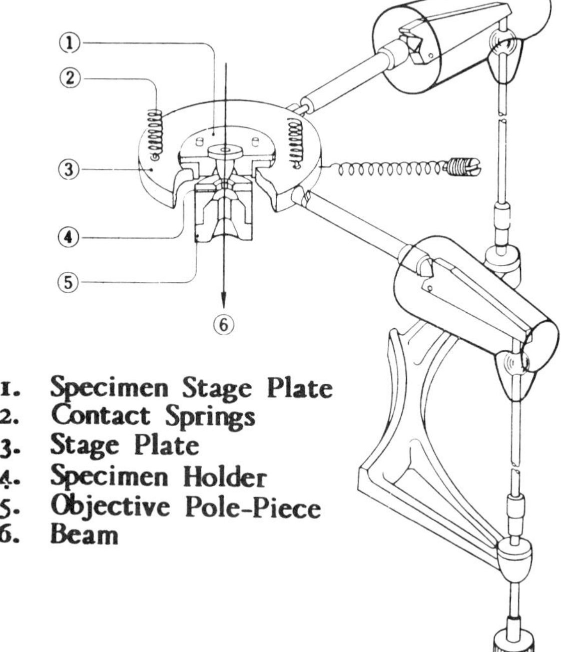

Figure 5-10. The specimen stage. (a) Stage and objective lens assembly. (b) Schematic diagram of the stage control unit. Courtesy of Hitachi, Ltd.

(Fig. 5-11). When the specimen is inserted, it is usually placed within the magnetic field of the objective lens (immersion objective). The specimen is placed on grids coated with a thin electron-transparent film (*see* Chap. 10). The specimen holder must make a good contact with the specimen grid and stage. Top-entry specimen holders are commonly used, but the recent trend is to use side-entry specimen holders. Both holders have their advantages and disadvantages. The top-entry types are simple in design, and the specimen height with these types can be adjusted to vary the focal length of the objective lens (*see* Chap. 4). Most top-entry systems can carry only one specimen at a time. However, a dialtype of multiple specimen holder

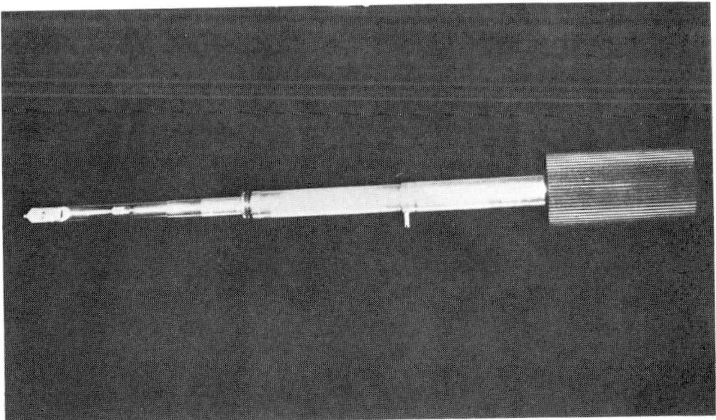

Figure 5-11. Specimen holders. (a) A top-entry specimen holder. (b) A side-entry specimen holder.

containing five to six cartridges is now available to overcome this problem. The side-entry system allows a simpler exchange of multiple specimens, and the focal length of the objective can be very short. The most outstanding feature of this system, however, is that a eucenteric goniometer tilting stage can be used with this, which is not possible with the top-entry system. The specimen holders are made of brass or copper.

Specimen Airlock

The specimen holder must be rapidly exchangeable from the outside of the column while the beam is still on so that the microscope will be ready to use immediately after the specimen has been inserted. This requires a specimen airlock, and different manufacturers use various means for this purpose.

Tilt Stages

Specimen stages are available for rotating and tilting the specimen with respect to the beam through known angles. The specimen can be tilted for stereomicroscopy, selected crystallography for revealing structures, and for obtaining different views of a complex structure. A desired orientation can be obtained either by rotating the specimen first and then tilting it, or by providing two orthogonal tilt directions that can be used in combination.

STEREOSCOPIC STAGES: With this stage, a specimen is tilted on either side of the axis through a small angle to yield three-dimensional information of the specimen (*see* Chap. 7).

GONIOMETER STAGES: These stages, either top-entry (Fig. 5-12a) or side-entry (Fig. 5-12b), allow the specimen to be tilted through a large angle (±60°) so that the structures are oriented parallel to the beam. A wide-angle double tilting stage allows different views of sections of complex biologic specimens that are especially useful in reconstruction of images. However, these stages need extreme care in their construction because tilting of a specimen produces an out of focus image. A loss of resolution and magnification usually occurs during this operation. These disadvantages can be overcome by the use of the more convenient side-entry *eucentric goniometer* or *axis-centered stages* (Fig. 5-12b). They allow wide-angle tilt (*see* Chap. 7)

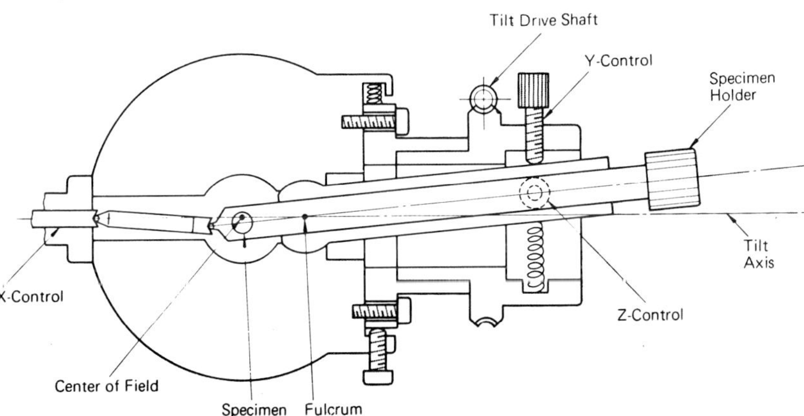

Figure 5-12. Top-entry and side-entry goniometer stages. (a) The top-entry goniometer stage assembly. (b) A cross-sectional diagram of the side-entry eucentric goniometer stage. Courtesy of Hitachi, Ltd.

plus the rotation of the specimen so that it can be tilted in two mutually perpendicular axes, and an image shift does not occur during the tilting operation. However, they are generally bulky, and some loss of resolution apparently still results.

The Camera

Although some TEMs are equipped with 35-mm cameras, most instruments contain plate cameras for permanent recording of electron images. Eighteen to twenty-four plates or sheet films of different sizes can be used in these cameras, and they are usually placed below the fluorescent screen. The recording system (Fig. 5-13) usually consists of a camera chamber located in the back with a lighttight metal box containing the casettes. A carrier mechanism (rack) is provided for drawing the casette to a position under the screen. After the plate is exposed, it is again carried by the carrier mechanism to a casette receiver, which is located in front. In most modern instruments, a serial number and the magnification are automatically printed on the plate. A 70-mm camera may be used in place of plates in some microscopes. The recording system is equipped with an airlock.

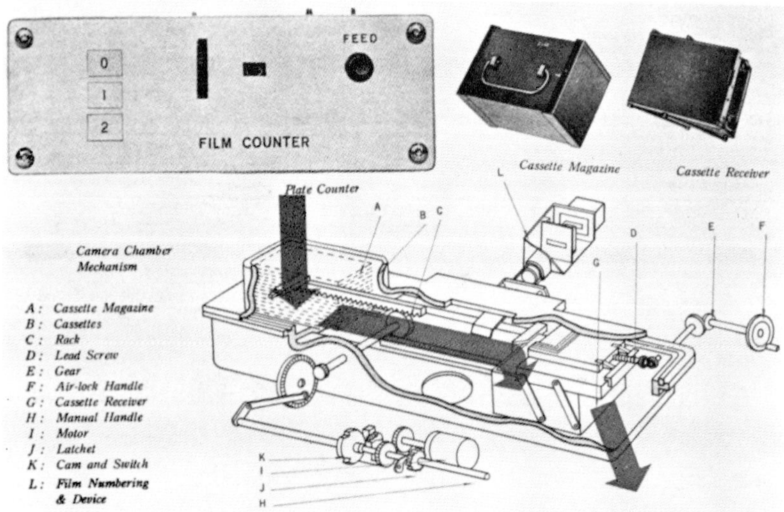

Figure 5-13. Schematic diagram of the image recording system. Courtesy of Hitachi, Ltd.

Plate or Film Reservoir

Most modern TEMs also provide a chamber for keeping a box loaded with casettes or films. Thus, the plates are kept under vacuum and the camera chamber can be reloaded with minimal pumping time. This reservoir is airlocked.

The Vacuum System

A high vacuum in the electron microscope column is needed for the following reasons: (1) unlike photons, electrons are charged particles, and at or near atmospheric pressure, the high-velocity electrons can be stopped upon striking the air molecules over a path of a few millimeters. This collision will scatter the beam and cause glare or reduced contrast in the image; (2) presence of gases causes high voltage breakdown between the cathode and anode, which results in electrical discharges. The electron beam then flickers or becomes unstable; (3) the filament life is substantially shortened in the presence of excessive gases, particularly oxygen; and (4) residual gas molecules cause rapid contamination of the specimen.

The vacuum is measured in terms of air pressure, generally in terms of mm Hg (preferably "torr," named after Torricelli, the discoverer of the barometer). A high vacuum of 10^{-4} to 10^{-6} torr is considered adequate for electron microscopy, and the "mean free path" of electrons in this vacuum is 2 to 3 m (the height of the column is approximately 1 m). However, it should be emphasized that the higher the vacuum, the better the result. The vacuum system of the microscope is continuously pumped to maintain the vacuum because gases enter the column continuously through minute leaks. Other gases, air, and water vapors also arise from the inside surfaces of the column and photographic materials. The vacuum system of the instrument contains a minimum of two pumps hooked in a series, a *mechanical rotary pump* (fore pump, backing pump) and a *diffusion pump.* The mechanical pump can evacuate to approximately 10^{-2} torr, and a diffusion pump that uses oil or Hg is used to achieve a high vacuum of 10^{-4} to 10^{-6} torr. A typical vacuum system is schematically presented in Figure 5-14.

The mechanical rotary pump (Fig. 5-15a) consists of an eccentric vane-mounted rotating cylinder immersed in oil in a cylindrical casing. As the eccentric cylinder rotates, it produces continuous

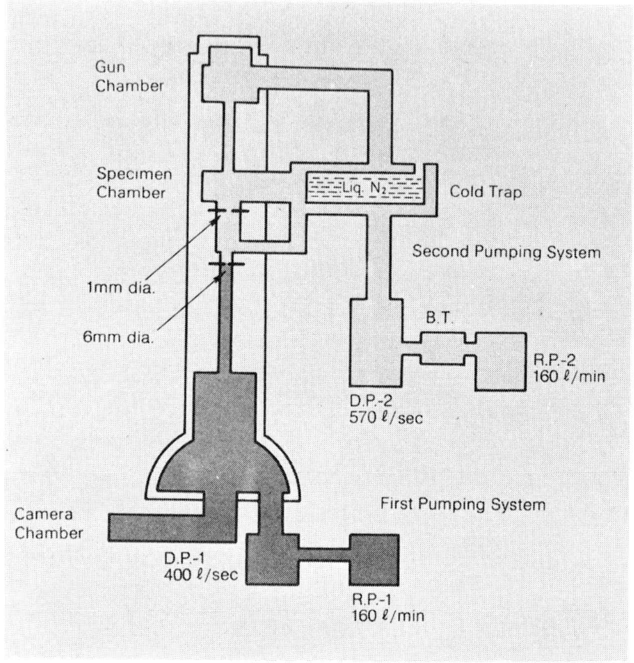

Figure 5-14. Diagrammatic representation of the vacuum system. Courtesy of Hitachi, Ltd.

streaming and compression of air molecules. The compressed gas confined to the outlet valve of the pump is forced out the exhaust pipe.

After the initial evacuation by the rotary pump, a diffusion pump (Fig. 5-15b) reduces the air pressure to 10^{-4} to 10^{-6} torr. The air molecules from the column entering the diffusion pump are trapped by jets of oil or mercury molecules directed downward. Oil diffusion pumps are commonly used in electron microscopes. Boiling of the oil (usually silicon oil) at the base of the pump creates high-speed jets of oil molecules, and the air molecules diffuse into these jets moving downward to the bottom of the pump. This establishes a pressure gradient that produces a pumping action. Hot oil molecules colliding with the cool wall (kept cold by running water) of the pump condense and return to the boiler. The backing pump removes the gas from the bottom of the diffusion pump.

Recently turbomolecular pumps that are mechanical and hydro-

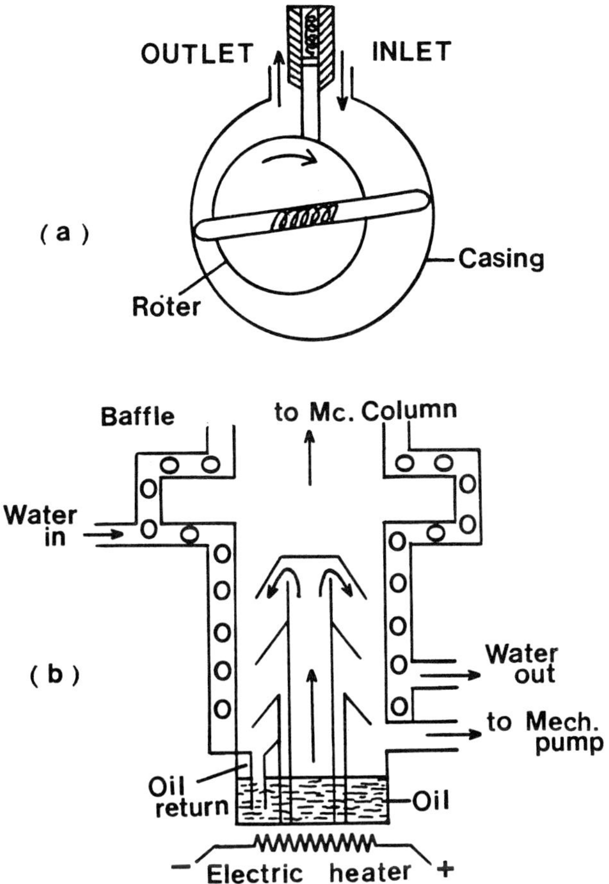

Figure 5-15. Vacuum pumps most commonly used in electron microscopy. (a) Mechanical rotary pump. (b) Oil diffusion pump. Adapted from G.A. Meek, *Practical Electron Microscopy for Biologists*, 2nd ed. Copyright C 1976. Reprinted by permission of John Wiley & Sons, Ltd., London.

carbon free have been introduced. They can attain a vacuum of 10^{-5} to 10^{-8} torr. The air molecules are driven out by moving and stationary slotted disks resembling the stages of a turbine. These pumps do not use oil, do not permit backstreaming, and are less contaminating than the oil diffusion pump.

Very sensitive vacuum gauges are placed in the electron microscope to measure the pressure in the column. A thermocouple gauge is used for measuring low vacuum (10^{-1} to 10^{-2} torr), and a penning gauge is employed for measuring high vacuum (10^{-3} to 10^{-6} torr).

For convenience, fully automatic vacuum systems are now standard in modern high-performance instruments. The complete sequence of vacuum system operation is carried by activating a single button in them. Electron microscopes are usually equipped with a vacuum failure switch, and the high voltage is automatically cut off when the vacuum pressure is low. It is very important that atmospheric air not enter the hot diffusion pump oil, which would cause the oil to decompose and contaminate the entire column by blowing the oil vapor into it. The failure of cooling water supply to the pump would also heavily contaminate the column. In modern instruments, the diffusion pump switch is operated by a water pressure switch, which activates a warning bell circuit and turns off the diffusion pump heater and high voltage supply. It is, therefore, important that the operator be familiar with the operation of the vacuum system, especially if it is not automatic, to avoid any accidents. The operation of the vacuum system varies from one manufacturer to the other, and the operator must follow the instruction manual of the instrument.

Electrical Power Supply

The electrical supplies are needed to provide (a) the HV to the electron gun for accelerating the electrons; (b) current to heat the filament; (c) current supplies for the electromagnetic lenses; and (d) current for various other circuits such as the vacuum system, stigmators, deflectors, bias voltage for cathode assembly, camera system, wobbler focusing, interlocks, safety devices, relay switches, and panel and room lightings.

The electronic circuits of modern electron microscopes are highly sophisticated and complex and usually require repairing by experts. The biologist will be better off using the service provided by the manufacturer for this purpose. Transistorized power supplies that operate at relatively low voltages have now virtually replaced the electronic valves. Highly stabilized power supplies are necessary for the HV, lenses, deflectors, and stigmators. A moderate degree of stabilization is needed for the oil diffusion pump and auxilliary systems. Fluctuation in line voltage supply (mains) can be stabilized by constant voltage transformers or motor-driven regulators. Regulators are also used for electronic circuits.

High Voltage Supply

The high tension supply in the electron microscope is a low current ($\approx$ 50 μA) HV type. It is negative to the cathode assembly and positive to the anode. Used to accelerate the electrons for penetrating the specimen, the HV has to be highly stabilized to only a few parts per million to minimize the effect of chromatic aberration for a high resolution (*see* Chap. 3). The entire HV generator is immersed in an insulating oil tank having no air bubbles or particulate matter to avoid discharge or arching.

Manufacturers use different methods for generating the HV. Usually it is generated by feeding AC mains to high tension circuits where it is first rectified to DC and stabilized. A high frequency oscillator controlled by the signal amplifier is then energized to provide a constant voltage output. This high frequency is applied to a cascade voltage multiplier that generates the HV and rectifies it. For stabilization of the HV to a few parts per million, it is also transferred to a separate measuring resistor in another oil tank for obtaining the feedback signal. This reduces the *noise* created by thermal movements in the oil or particles released from electrical components in the main tank.

Manufacturers now provide an accelerating potential range of 20 to 125 KV (*variable accelerating potential*), and the operator can select the desired HV at regular steps. With the accelerating voltage range of 40 to 100 KV, the effective wavelength of the electron beam is insufficient to affect the resolution of the microscope (*see* de Broglie's equation, Chap. 1). The choice of the accelerating potential, therefore, depends on the requirement for penetration and contrast. The scattering of electrons by an object of a given thickness decreases with increasing voltage (*see* Chap. 4). Hence, a beam of higher voltage can be used to visualize the internal structures of a relatively thick specimen. Conversely, the contrast of a thin specimen is increased when a lower accelerating potential is used. Therefore, a 1,000Å particle is best examined at a higher voltage, e.g. 75 to 125 KV, and a thin section at a lower KV, e.g., 50 KV. Excessive energy loss of incident electrons occurs when a thick specimen is examined with a lower accelerating potential that results in high chromatic aberration and poor resolution (*see* Chap. 3). Thus, it is better to accept some loss of contrast to obtain a better penetration at a higher KV.

Filament Heater Current Supply

Unlike the HV supply, this current supply is of a high current and low voltage type. The tungsten filament requires about 2 to 3 A and the current needs a moderate degree of stabilization. Either an AC or a DC supply can be used although the recent trend is to use a DC supply for the highest stability. The self-biasing resistance is housed in the HV tank. When the bias voltage is decreased, the beam current is increased, and the filament is usually saturated at a higher temperature, resulting in a higher beam brightness. The filament heater current and bias controls are placed on the desk panel of the microscope. There is an electron beam current meter as a guide for setting these controls. This meter also indicates a *dark current* (standing current) due to a very small current passing through high resistance of the bleeder chain across the HV supply whenever the HV is switched on (before the filament current is applied). The beam current due to electron emission then rises as the filament is heated. The power supply to the cathode assembly in the electron gun is carried by insulated leads in the core of the HV cable.

Lens Current Supply

A highly stabilized power supply is necessary for energizing the individual lenses, and it is now mostly transistorized. A stability of only a few parts per million is required for the objective lens (one-half of that of the HV supply) for better resolution because a fluctuation in objective lens current gives rise to chromatic aberration (*see* Chap. 3). The lens coils in some microscopes are also cooled by means of a water supply. Other lenses require a moderately stable power supply.

The programming of the lens current supply is usually built into the instrument to have a correct lens strength combination especially for reducing the distortion effect. When the magnification is varied, the second condenser automatically adjusts its strength to maintain a near constant intensity as does the objective. There are many different controls on the instrument panel, e.g. intensity, focus, and magnification and most microscopes now provide a direct magnification readout on the panel.

Currents for Other Circuits

Besides the stigmators, which need a highly stabilized power supply, a moderately stabilized current supply is used for other auxiliary circuits. Most of them are now operated with a DC supply because AC supplies usually create a fluctuating magnetic field that can cause movement of the beam. There are circuit monitoring points in the instrument that should be periodically checked, particularly the main reference circuit.

SELECTED BIBLIOGRAPHY

Agar, A.W., Alderson, R.H., and Chescoe, D.: *Principles and Practice of Electron Microscope Operation.* In Glauert, A. (Ed.): *Practical Methods in Electron Microscopy*, vol. 2. Amsterdam, North-Holland, 1974.

Chapman, S.K.: *Understanding and Optimising Electron Microscope Performance.* A Perkin-Elmer EM Publication.

Hall, C.E.: *Introduction to Electron Microscopy*, 2nd ed. New York, McGraw-Hill, 1966.

Meek, G.A.: *Practical Electron Microscopy for Biologists*, 2nd ed. London, Wiley, 1976.

Sjöstrand, F.S.: *Electron Microscopy of Cells and Tissues*, Vol. 1, *Instrumentation and Techniques*. New York, Academic Press, 1967.

Wischnitzer, S.: *Introduction to Electron Microscopy*, 3rd ed. New York, Pergamon, 1981.

Chapter 6

COMMERCIAL TRANSMISSION ELECTRON MICROSCOPES AND THEIR INSTALLATION

VARIOUS LARGE INTERNATIONALLY reputable firms with vast capital resources, established departments of electron optics, electronics, general instrumentation, and vacuum technology manufacture numerous models of transmission electron microscopes (TEMs). These instruments are highly complex in design and, among the scientific instruments, are generally a good buy for their value. Unaware of the basic design, however, a prospective buyer can be easily blinded by the accessories and frills offered by the manufacturer.

The leading manufacturers at present are: in Japan—Hitachi Ltd. (microscope model prefix—Hitachi HS, HU, H), and Japan Electron Optical Laboratory (JEOL) (model prefix—JEM); in England—Associated Electrical Industries (AEI) (model prefix—AEI-EM); in the Netherlands—N.V. Philips Gloeilampenfabrieken (model prefix—Philip-EM); and in Federal Republic of Germany—Siemens (model prefix—Siemens Elmiskop), and Carl Zeiss (model prefix—Zeiss EM). In the U.S.A., the Radio Corporation of America, which manufactured TEMs since 1941, has been out of this market since the late 1960s. The manufacturers offer TEM models that may be classified as medium performance and high performance instruments. Their prices are rising every year due to inflation, and a high performance TEM with some essential accessories may cost as much as $160,000.00, which includes delivery, installation, instruction for operation, and a one-year warranty. In the present state of innovative art, the modern instruments are sold so that they can be expanded to a complete analytic system at any time without disturbing their integrity.

It should be pointed out that a used instrument can be an extremely good buy, especially when it has been well maintained. This should be explored when the prospective buyer has limited funding. Manufacturers are constantly changing the models for competition and prestige, and trade-in instruments can be purchased at a reasonable price.

MEDIUM-PERFORMANCE MICROSCOPES

These instruments can satisfy the need of many biologists, especially the cell biologists, at a much lower price than the high-performance equipment. The resolution range of these instruments is 5 to 10Å, and the magnification varies from 50,000× to 200,000×. Most instruments of this type have stigmators, anticontaminators, and vacuum automation. A few examples of this type are Hitachi HS 9, and HU-11 series, Philips EM 201, AEI-EM-6, JEM-7, Siemens Elmiskop 101, and Zeiss EM 109.

HIGH-PERFORMANCE INSTRUMENTS

These offer an ultrahigh resolution of as low as 2 to 3Å with a magnification as high as 500,000× and are possibly the ultimate in transmission electron microscopy. They offer a fully automatic fail-safe vacuum system, accurate specimen manipulation, extremely stable electrical supplies, and a variety of convenient accessories. Attachments for scanning electron microscopy, scanning transmission electron microscopy, x-ray microanalysis (Fig. 5 1), and image intensifiers are also available. However, these are not required by most users. Some examples of this type are Hitachi H-600, JEM 100C, Philips EM 401, Siemens Elmiskop 102, AEI-EM-8, and Zeiss EM 10C.

Transmission Electron Microscope Features Required By The Biologist

It is emphasized that an ultrahigh resolution guaranteed by the manufacturer is either rarely achieved or required in biologic electron microscopy. The biologist should choose the instrument for necessity rather than for laboratory prestige. The following features are usually required.

Double Condenser System

The advantages of this system are described in Chapter 5. A double condenser is essential for an instrumental magnification above 50,000×. The second condenser lens should have an electric stigmator. A beam-shifting facility may be required by some biologists.

Prealigned Lenses

Factory preset lenses are convenient to a biologist for alignment of the microscope.

Anticontaminator

This is essential for high-resolution work (*see* Chap. 8). A cold finger or cryopump is the most efficient anticontaminator.

Tilting Stages

These are worthwhile features that are becoming increasingly necessary in biologic electron microscopy. A biologist may need stereo operation or wide-angle double tilting. Eucentric side-entry goniometer stages are much more simple in operation (*see* Chap. 5) and enable the specimen to be rotated as well as tilted at a wide angle. The angle in degrees is usually indicated on a dial beside the stage.

Low Magnification Scan

A desirable feature, scanning at a low magnification allows a rapid survey of the specimen. A suitable area can then be selected for examination at higher magnification.

Wobbler Focusing

This is a very important aid for focusing the image at a

magnification below 10,000× because the electron image is difficult to focus at a low magnification.

Automatic Vacuum System

A fully automatic vacuum system is quite convenient to the biologist, but it must be extremely reliable and fail-safe. All high-performance microscopes now have this system.

Multiple Specimen Holder

This is extremely convenient and time saving. Both side-entry and top-entry holders of this type are available.

Four-Lens Imaging System

This is required for high instrument magnification and is convenient for diffraction work.

Variable Projector Lens Strength

This good feature is helpful in reducing distortion at low magnification. Some manufacturers provide projector turret controls for selecting the proper polepiece.

Variable Accelerating Potential

This is now standard in all high-performance instruments. A lower KV, e.g. 50 KV, is usually used for thin specimens, and a higher KV, e.g. 75 KV to 125 KV, is employed for thick specimens, particularly when wide-angle tilting is desired.

Electrical Shutter

This is essential for photographic recording of the image. It is standard in all high-performance instruments and in most medium-performance microscopes.

Other Features

The following features are generally nonessential for biologic electron microscopy. These include selected area diffraction, x-ray microanalysis, and serial-section specimen holders. A luxury item, *zoom focusing*, incorporates a stepping system that adjusts objective lens strength to appropriate values each time the current is varied. Thus, there is little change in focus or image brightness. An *image intensifier* is used to increase the brightness of the image electronically and display on a cathode ray tube. Thus, it is good for teaching and demonstration to a group. It can be employed to examine beam-sensitive specimens at an extremely low beam intensity. Therefore, the specimen can be observed at a very high magnification with little loss of contrast or brightness. The video output signal can also be directly fed to a videotape recorder. This device is expensive and nonessential.

The necessity for other frills such as interchangeable cameras, automatic water temperature control, and automatic adjustment of the intensity of panel lighting is only marginal.

INSTALLATION

Before purchasing a TEM, consideration should be given to its housing, for which a representative of the manufacturer should be consulted.

The purchase price usually includes delivery of the instrument, installation, and instruction for operating the microscope, its maintenance, and a one-year warranty. All wirings and plumbings should be completed before the instrument is delivered. An EM is a delicate, expensive instrument, but once properly installed, it is remarkably durable, robust, and reliable. The following factors should be considered.

The Microscope Room

Most TEM facilities now consist of a single unit with four to five rooms. A room should be provided for the microscope itself, and it should not have other apparatus such as a vacuum evaporator or photographic equipment. A revolving door between the microscope and the adjacent rooms is desirable. The microscope room should be clean, air conditioned, well ventilated, and dust-free. It must be capable of being completely blacked out and easily cleanable. The floor should be plain and preferably white, and it must be strong to withstand the heavy instrument. There must be adequate room around the microscope for servicing, particularly in the back of the column. The power supplies should preferably be placed in an adjoining room because they give rise to large magnetic fields. The rotary pump produces vibration, and it should also be isolated and placed in this room. One adjoining room should preferably be available for photographic equipment and developing and another for specimen preparation. The unit should also have an office for the microscopist.

Vibration

Unlike the older generation of TEMs that were quite sensitive to vibration and needed a concrete floor in the basement for their installation, the new generation instruments are not as sensitive to mechanical vibration. Although a basement is undoubtedly ideal, the modern microscopes can be installed on not so ideal flooring. The microscope, however, should not be installed close to an elevator. Excessive mechanical interference due to vibration should always be avoided.

External Magnetic Field

Electromagnetic TEMs are sensitive to external magnetic fields. Tests should be made to detect any stray magnetic field before the microscope is installed. Electric transformers, motors, and other equipment in neighboring laboratories usually interfere with the microscope operation.

Electric and Water Supplies

It is important that the microscope have a separate, main electrical supply, and a 220-volt line is preferable over a 110-volt line. Since transformers in the power supplies produce magnetic disturbances, these supplies should preferably be kept in the adjoining room.

A reliable cold water supply with a constant pressure is necessary for the oil diffusion pump and lenses. It is better to place a filter in the line, and the filter should be replaced before it is completely clogged. To avoid corrosion, the water should not be too soft or too hard. Many electron microscope laboratories use a closed-circuit water system. The cooling water is continuously recycled with a pump and cooled by refrigeration.

Radiation Hazards

The sources of x-rays from the TEM are from the electron gun, the specimen, condenser and objective apertures, and the fluorescent screen. The manufacturers use proper shieldings to protect the operator against this radiation hazard. Lead or other dense metal is used around the gun, and thick lead glasses are used for viewing windows. The microscope should be periodically checked with a Geiger counter while it is operating at the maximum high voltage and beam current. The instrument should never be operated with the shieldings removed. The electron microscopists should wear a radiation badge, and a badge should also be strapped to the instrument.

Repair Service

Most manufacturers provide adequate instruction for operating and maintaining the instrument. A biologist can easily replace a burned filament, clean the apertures, and conduct minor repairs. With column jacks standard in all modern microscopes, he may also be able to clean the polepieces. However, for repairing the electronic circuits and for other complex repairs, it is best to seek service provided by the manufacturer. The contract service charges for these instruments are increasing every year. With all its complexity and sophisticated design, the TEM is quite reliable. In fact, it is so reliable that care should be taken not to provide excessive maintenance.

SELECTED BIBLIOGRAPHY

Agar, A.W., Alderson, R.H., and Chescoe, D.: *Principles and Practice of Electron Microscope Operation*. In Glauert, A. (Ed.): *Practical Methods in Electron Microscopy*, vol. 2. Amsterdam, North Holland, 1974.

Meek. G.A.: *Practical Electron Microscopy for Biologists*, 2nd ed. London, Wiley, 1976.

Chapter 7

OPERATION OF THE TRANSMISSION ELECTRON MICROSCOPE

THE DESIGN AND OPERATION of each model of commercial transmission electron microscopes (TEMs) vary considerably, and it is not possible to provide a precise method for operating these microscopes. The operation manual accompanying the instrument gives detailed instructions for starting the microscope from the cold, step-by-step alignment, correction of astigmatism, photography, and shutting down procedures. The modern instruments with a considerable amount of automation are now easy to operate because much of the alignment is preset. It is recommended that the operator be familiar with all the controls on the instrument and its desk panel and their function before attempting to operate the microscope. The operation manual should be consulted for locating the position of these controls.

ALIGNMENT

The conditions achieved by proper alignment of the microscope have been described in Chapter 5. Misalignment can interfere with resolution, and an instability of high voltage (HV) or lens current causes a movement of the image, which increases with increased misalignment. Alignment of the column is necessary to minimize the effects of circuit instability, which causes a loss of resolution away from the axis. Most of the alignments are fairly stable, and realignment is rarely required until the filament is replaced. Only a routine quick check is usually necessary to determine the accuracy. Some of the illumination controls, however, need frequent adjustments, especially the illumination traverse controls. Since the instrumental designs differ, the orders in which the various microscopes are aligned may be different. The operator should follow the alignment steps outlined in the instruction manual.

A stepwise alignment starting with the gun, condenser, and the imaging system is usually done. The highest possible KV should be used while retaining sufficient contrast. After adequate vacuum has been obtained, the required HV is switched on, and the beam current meter is checked for residual dark current (*see* Chap. 5). The operator should wait until this current falls below 5 μA. The filament current is then applied, and the filament is saturated (as indicated by the leveling of the beam current meter). Once saturation is reached, the filament current control should not be advanced further. If the beam current flickers, it usually indicates a small corona discharge in the gun due to a poor vacuum or some dust particles.

The gun should be operated with a small degree of bias to obtain a small source size. An unbiased gun produces a large spot size. The beam current is a monitor of the extent of specimen damage, and a minimal beam current should be used, provided the illumination is bright enough for operating comfortably. After locating the beam, the gun is centered. The condenser illumination and crossover are then centered with the gun and objective lens. The intermediate and projector lenses are now centered with respect to the first and second condensers and the objective. The second condenser aperture is then inserted and centered on the crossover. The smallest possible condenser aperture along with the smallest spot size should be used while retaining sufficient brightness of illumination. The gun is now centered with respect to all lenses. After adjusting the second condenser lens (C_2) for crossover, the filament is desaturated (under heated) to obtain the image of the filament and an even halo around the central *draw lines* (Fig. 7-1). The filament is then resaturated.

Correction of Condenser Astigmatism

The astigmatism in the C_2 affects the brightness of the illumination and can cause a slight loss of resolution at the highest magnification. It should be corrected by one of the following methods: (a) with the condenser aperture in place but the C_2

Figure 7-1. An underheated filament with halo, the bias shield, around the central wire-drawing lines in the tip of the tungsten filament.

stigmator off, the direction of the astigmatism is determined by adjusting the C_2 to near crossover. The corrector is switched on and the amplitude (strength) control is increased until the image becomes elongated. It is then rotated at right angles to the original direction by the azimuth (direction) control. Finally, the strength is reduced until the image of the condenser aperture is circular. The spot will not be circular if the aperture is contaminated; (b) the C_2 is focused at a low magnification and the filament is underheated until the *draw lines* of the filament tip are focused. Deastigmation is done by manipulating the C_2 stigmator controls until the draw lines appear the sharpest (Fig. 7-1). The filament is then resaturated and the circularity of the condenser aperture image is checked; and (c) the C_2 is focused at a low magnification with a low beam, the C_2 aperture is withdrawn, and the caustic pattern formed by the lens is observed. The lens is deastigmated by adjusting the corrector until the caustic pattern becomes a three-pointed star (Fig. 7-2). This is the most accurate method, but this pattern cannot be obtained if the C_2 polepiece is contaminated or damaged.

Voltage Alignment or Centration

After the gun, condenser, and imaging systems have been aligned, a specimen (a holey film is most suitable) is inserted, and the voltage center is located. This is done by varying the accelerating potential (by activating the voltage alignment switch) while the objective lens current is fixed. A focused image will contract and pulsate around a common point in the image plane called *voltage center* (Fig. 7-3a).

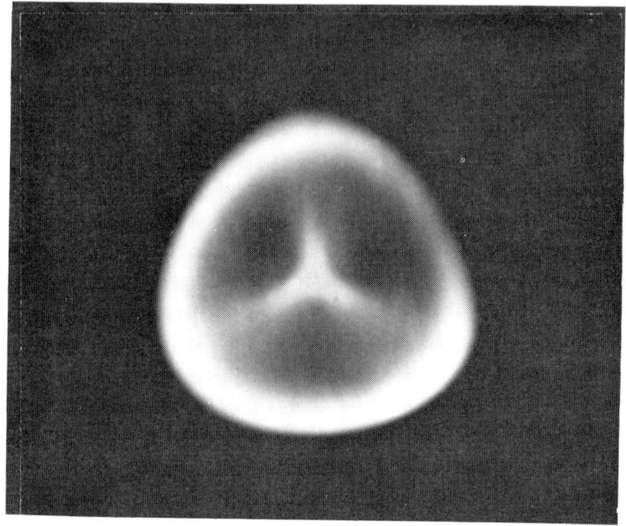

Figure 7-2. The three-legged caustic figure with second condenser lens astigmatism corrected.

For a certain change in the HV, the movement of an image point over a distance will increase with distance from the voltage center. Therefore, the image quality due to HV fluctuation will be least affected if the voltage center is at the center of the screen. The voltage center defines one of the optical axes of the electron microscope. An identical effect is observed when the objective lens current is varied while the accelerating potential is fixed. This is the *current center*, the other optical axis. These two centers usually do not coincide. Since the high voltage is more susceptible to random fluctuation, the voltage alignment method is generally used. The voltage alignment (by condenser tilt control) is completed when the image pulse is brought to the center of the screen and the movement of a point in the image is the least (Fig. 7-3b). The condenser traverse should be kept in alignment during this operation. The voltage alignment control should now be switched off.

Centering the Objective Aperture

A diffraction spot is obtained by demagnifying the intermediate lens and a clean objective aperture is inserted and centered over the diffraction spot.

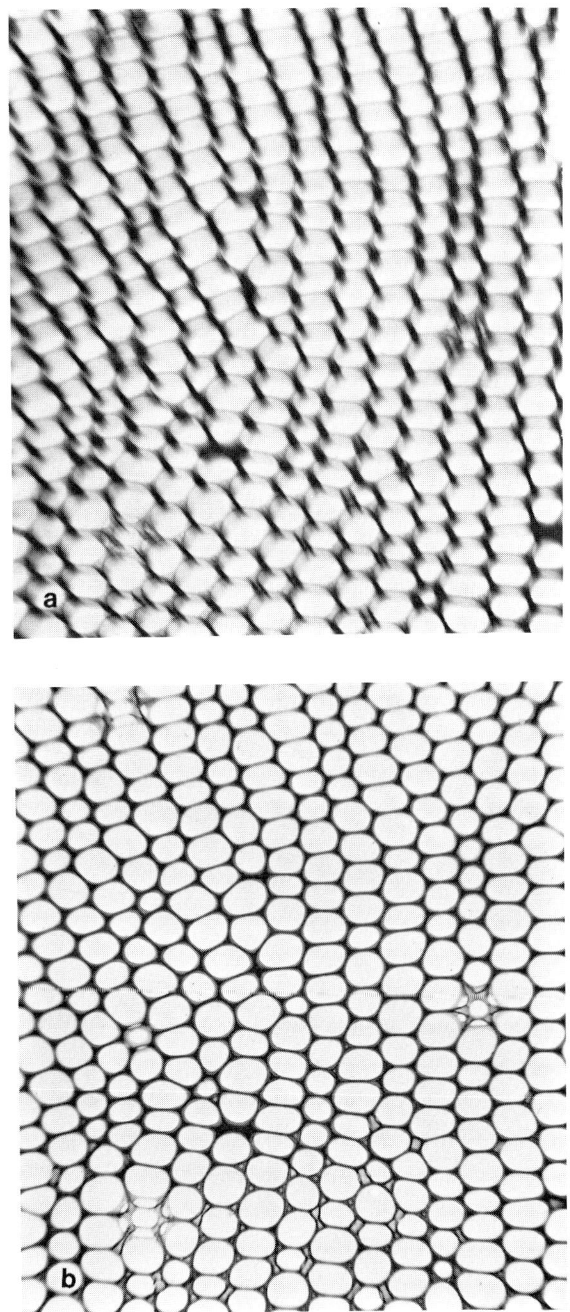

Figure 7-3. High voltage centration. (a) Location of high voltage center. (b) High voltage properly centered. Courtesy of Hitachi, Ltd.

Correction of Objective Lens Astigmatism

Astigmatism of the objective lens severely limits the resolution, and it must be corrected for achieving the optimal resolution, (*see* Chap. 3) and for working at magnifications higher than 40,000×. This is a correctable instrumental defect, and electron microscopists usually check the residual astigmatism before high resolution photography. The effect of astigmatism is best observed with a small, nearly round hole in a carbon film (holey film). The entire hole should be visible with the binocular.

The objective lens current is adjusted to obtain a slightly overfocus Fresnel fringe so that the fringe just touches the edge of the hole. The fringe will differ in magnitude (width) or stick to the edge at two points if the astigmatism is present (Fig. 7-4). With the stigmator off, the direction of the astigmatism is noted (Fig. 7-4a). When the corrector is now switched on with amplitude control at full strength, the fringe takes up a new direction (Fig. 7-4b). The azimuth control is then rotated to bring the overfocus fringe at right angles to the direction first observed with the stigmator off (Fig. 7-4c). The strength is then reduced to obtain a uniform fringe (Fig. 7-4d). The correction is usually done at a magnification twice the working magnification regularly used. The objective aperture can introduce astigmatism if it is contaminated or improperly centered.

Locating the Beam

After switching on the microscope or after inserting a new specimen, the operator occasionally may not see any illumination on the screen. This may be due to the following: (a) the HV may not be on and the filament may be burned (check the beam current meter); (b) the HV supply might have been automatically cut off due to a vacuum leak; (c) the specimen holder may not have been properly inserted or the specimen exchange device may be blocking the beam; (d) the C_2 may not be properly focused; (e) the condenser and objective apertures may be obstructing the beam; (f) the magnification may be too high, resulting in a very low intensity, and the illumination may not be noticed; (g) a grid bar could be obstructing the beam; and (h) the stage traverse control may be at one extreme end.

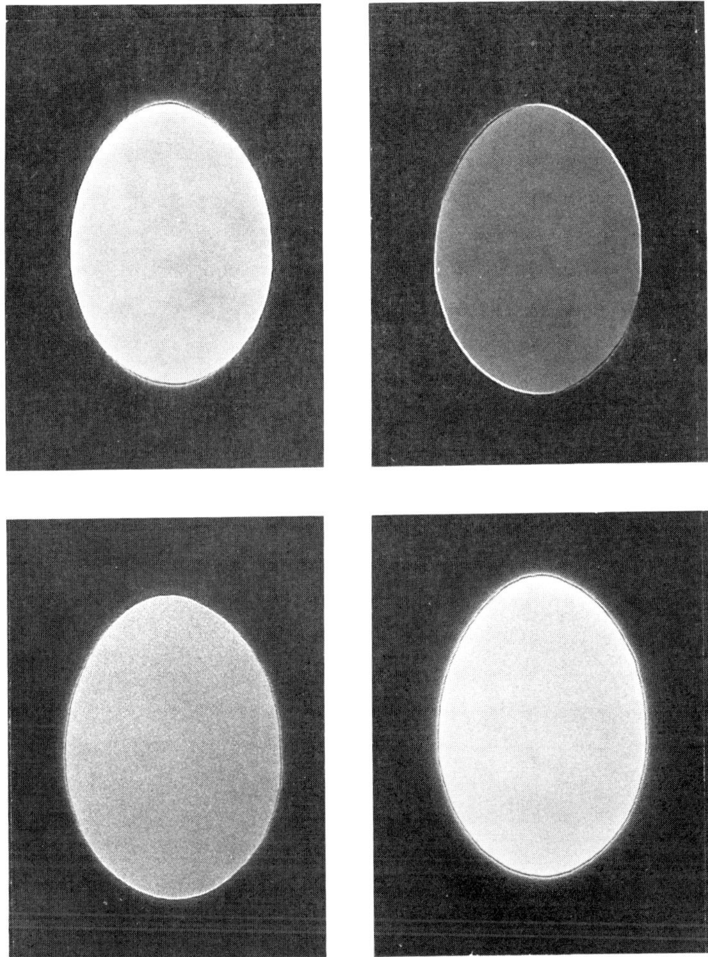

Figure 7-4. Astigmatism correction of the objective lens. (a) Slightly overfocused Fresnel fringe showing the presence and direction of astigmatism (stigmator switched off). (b) With stigmator switched on and at full strength, the direction is changed and there is increased astigmatism. (c) The direction of astigmatism rotated 90° to the original direction first observed. (d) Stigmator strength reduced and astigmatism corrected by obtaining a uniform fringe around the hole. (×95,000)

These points should be checked before moving a lot of controls that could cause misalignment and make it more difficult in finding the beam later.

Focusing the Image

When the magnification is changed by varying the strength of the intermediate lens, the image appears blurred, and the strength of the objective lens has to be adjusted to focus the final image. The image contrast is always minimal at the true focus, and the microscopist has to choose the right image for photography. An image in perfect focus has the highest resolution but may have a loss of some details due to lack of contrast. On the other hand, an image at a slightly underfocus setting has enhanced contrast and reinforcement of some details, but other smaller details may be missing due to a loss of resolution. Experienced electron microscopists generally like to photograph the electron image at a slightly underfocus setting. The focusing must be accurate at higher magnifications, and the viewing binocular should always be used for focusing the image before photography. The following methods are used as focusing aids.

Wobbler Focusing

Electron images are more difficult to focus at low magnifications than at high magnifications. For magnifications of 10,000× or below, wobbler focusing introduced by le Poole is the best method. A wobbler is now mounted above the objective lens of all high-performance instruments. When the beam is tilted back and forth with respect to objective lens axis by energizing the wobbler circuits, a movement (shift) of the image occurs. This gives rise to blurring if the objective is out of focus (Fig. 7-5). If the objective is in exact focus, no image shift and blurring occur. This method is not sensitive for use at a high magnification.

Minimum Contrast Focusing

In this method, the objective aperture is withdrawn, the objective lens current is varied, and the change of contrast is observed as the objective passes through the focus. The effect is striking with a holey carbon film. The contrast of a small thin specimen disappears

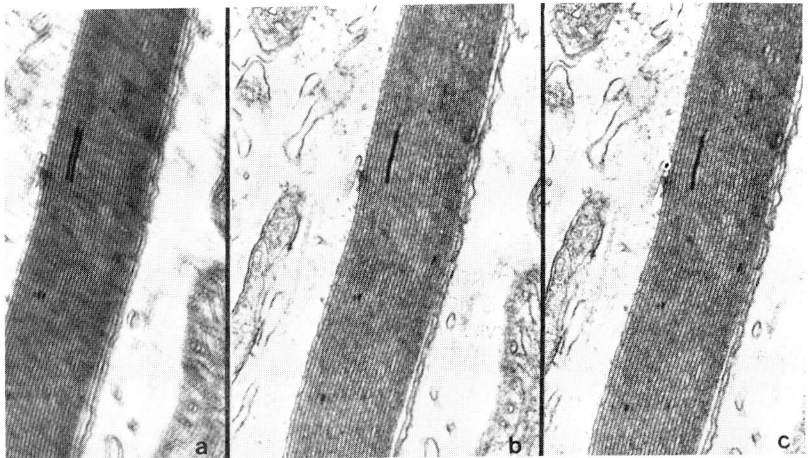

Figure 7-5. Wobbler focusing. (a) Out-of-focus, wobbler on. (b) In-focus, wobbler on. (c) In-focus, wobbler off. Courtesy of Hitachi, Ltd.

almost completely at true focus. The objective aperture is reinserted at this point and a photograph is taken. The method is fairly accurate at low magnifications.

Maximum Contrast Method

The maximum contrast is always a little under true focus (*see* Chap. 4). The method can be used for focusing at higher magnification and resolution. With the objective aperture in place, the magnification is increased. The objective lens current is then varied to observe the point of maximum contrast. Oncc this point is accurately judged after some practice, the number of fine-focus click stops of the objective lens between maximum contrast and true focus is determined. A correct amount of over- or underfocusing can then be applied for recording an image in focus.

Fresnel Fringe Method

This is the most accurate method especially at a very high magnification. The fringes are most conspicuous in a holey film (*see* Chap. 3) or around sharp particles (such as dirt or latex particles). The image of the edge is observed, and the objective fine focus is adjusted until there is no fringe or bright line around the edge. Although some electron microscopists place their specimens on a holey carbon film for this purpose, this is not possible with most specimens. If a small hole or a sharp particle is not present in the required field to be photographed, a through-focal series of micrographs should be rapidly taken. The technique is laborious and may be disappointing at times. Some experienced microscopists use the characteristic changes in the background structure of a carbon film for judging the true focus.

Taking the Photograph

The operator should follow the instruction manual for exposing the photographic plate or film. The illumination should be adjusted to appropriate brightness (by intensity meter) to expose the plates for minimizing graininess. The magnification selected should usually represent a large area of the specimen. Since a negative can be enlarged up to 20×, too high an optical magnification is rarely needed for recording the image. Once the area to be photographed is located, the plate is drawn under the screen. The image is checked for any drift, focused, the exposure meter is adjusted for proper intensity and time, and photographed. The plate is drawn into a receiver after it has been exposed. The photographic technique is described in Chapter 9.

Stereoscopy

The large depth of field and focus of the TEM successfully allow sharp in-depth views of the images of a specimen, and information on the third dimension of the object can be obtained (*see* Chap. 5). In the stereo operation, a specimen is first focused and photographed. The specimen is then tilted through an appropriate small angle (6 to 10°), refocused, and photographed again. When the two photographs are superimposed either physically if they are transparencies, or viewed with a stereoscope, a three-dimensional image is observed (Fig. 7-6). Stereoscopy in a high voltage electron microscope (*see* Chap.16) is particularly useful for observing three-dimensional images of a thick specimen.

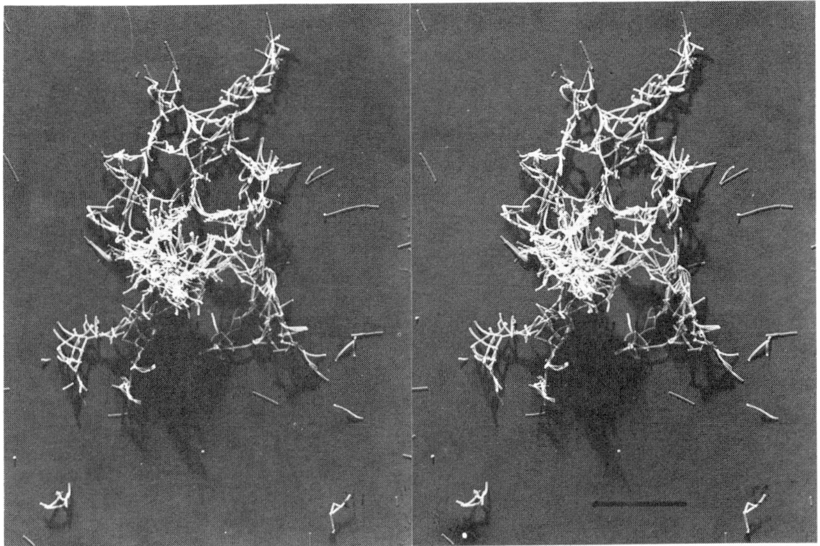

Figure 7-6. A stereo pair of shadowed tobacco mosaic virus particles. Bar = 2μm. Courtesy of Dr. R.L. Steere.

Wide-Angle Double Tilting

This method (*see* Chap. 5) is particularly useful for sections of biologic specimens with complex membrane structures. Different views of these complex structures can be obtained after wide-angle double tilting ($\pm 60°$) at several settings for a better interpretation of the image (Fig. 7-7). A tilting operation causes the image to move out of focus, and the image must be refocused after tilting. A change in magnification also occurs after a wide-angle tilting, and this must be compensated during photographic enlargement. An eccentric side-entry goniometer stage facilitates this operation. With these stages, the tilt axis intersects the electron optical axis, and an image shift does not occur. The tilting operation in most modern instruments is now motorized, and foot switches are used for this purpose.

Electron Diffraction

When a coherent beam of electrons strikes a crystalline specimen or a specimen with periodic structure, the electrons are preferentially

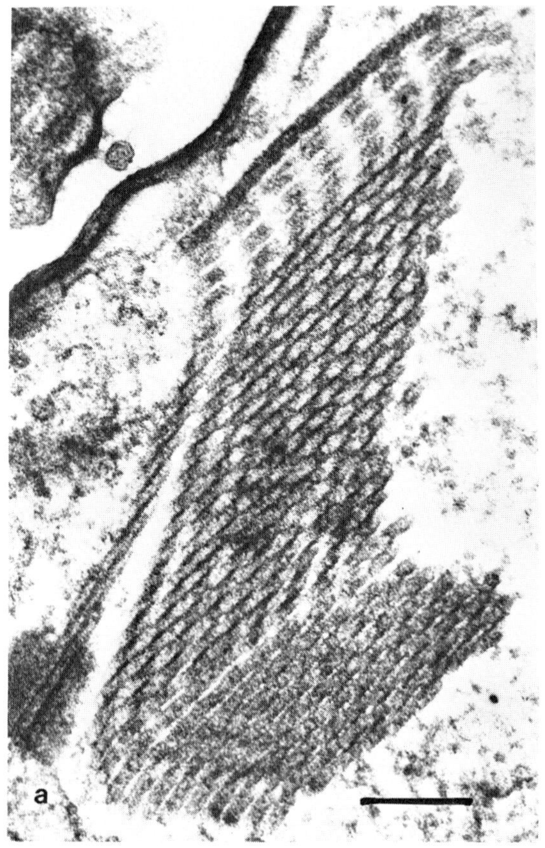

Figure 7-7. Wide-angle double tilting. (a) Microtubules in the mouth region of Condylostoma before goniometer tilting.

scattered into discrete rays by being reflected at the lattice structure. This is called *electron diffraction.* An amorphous object does not produce electron diffraction patterns. The diffraction angle can be calculated by Bragg's law:

$$2d \sin \theta = n\lambda$$

where d = distance between atomic planes (crystal lattice spacing); θ = diffraction angle; and n = an integer (whole number).

The final diffraction pattern of the specimen can be imaged as a

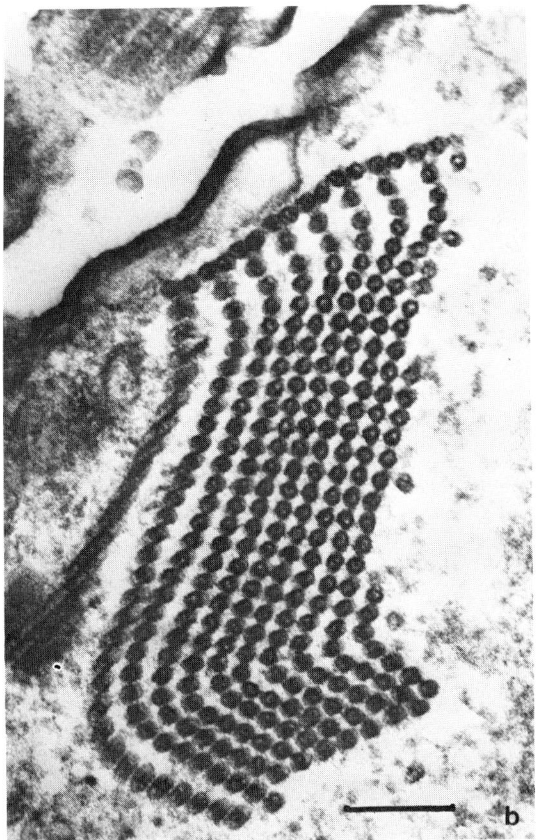

Figure 7-7. (b) The same specimen tilted 22° in the direction of the membranes. Bar = 100 nm. Courtesy of Siemens.

pattern of spots or rings of electrons (Fig. 7-8), and the dimensions and spacings of the spots can be used to calculate the lattice dimension of the crystalline object. Crystalline substances, such as metals, produce diffraction patterns, which are very useful in identifying them. Biologic specimens generally do not have this property.

A diffraction pattern can be imaged on the screen without any lens, but the pattern is very small, and it is usually magnified by the imaging system. The procedure for obtaining an electron diffraction

Figure 7-8. Diffraction patterns. (a) Iron cobalt alloy. (b) Aluminum. Courtesy of Mr. M.E. Taylor, Jr.

pattern is outlined in the operation manual. Briefly, a suitable specimen is selected, the area selector diaphragm and blades are inserted and focused by the intermediate lens. The specimen is then sharply focused by adjusting the objective lens current. The C_2 is adjusted to obtain a good illumination, the objective aperture is withdrawn and the diffraction pattern photographed.

Record Keeping

It is very important that the operator keep a record of the microscope operation, preferably in a hard-bound notebook. The microscopist should record the actual filament burning time (to keep track of filament life), the description of the specimen examined, the magnification at which it was photographed (that includes projector polepiece and intermediate lens current), and the KV at which the microscope was operated. When a plate is exposed, its serial number must be noted and identified. The magnification and serial number are now automatically printed on the plate during exposure in all high-performance instruments. The operator should also record any problems encountered with the instrument during the operation period. A record of service calls and repairs done to the microscope should also be kept. All negatives should be stored permanently for future references. The micrographs should be identified and the total magnification recorded on the back. Some interesting specimens may be saved although they are usually damaged after exposure to the beam.

Factors Affecting the Image Quality

Misinterpretation of electron micrographs taken at very high magnifications can usually occur, especially for interpreting finer structures (5 to 20Å). The following factors can affect the quality of the image.

Focusing

Defocusing of the image can introduce granularity that would be visible at a high magnification. Although the effect of defocusing is most striking in a holey film (due to Fresnel fringes), the structure of a carbon film is a good indicator of this effect. In the in-focus condition, the carbon film has very little structure. The size of the grain increases as the underfocusing is increased, and the contrast of these grains is reversed in the overfocus side (Fig. 3-9).

Astigmatism

This is a very common defect in the image because astigmatism cannot be perfectly compensated, and the objective aperture gets contaminated. The effect is pronounced at the edge of a hole in a carbon film. The overall effect of astigmatism on the image is a directional striated structure seen due to round objects being imaged as lines in one of the astigmatic planes.

Specimen Drift

This causes a uniform blurring and is commonly caused by thermal expansion due to heating of the specimen by the beam. It may also be due to a broken support film, heat absorption by large particles, and buckled grids making poor thermal contact with the specimen and specimen holder. A large drift is unlikely to be passed unnoticed in the binocular, but a very small drift may not be detected on the screen. Drift is minimized if the specimen grid makes a good thermal contact with specimen holder. It can also be minimized by scanning the specimen at a low magnification with a low beam intensity. A *curing* with a low-intensity beam is particularly needed to stabilize resin-embedded thin sections placed on an uncoated grid.

Mechanical Vibration

This gives rise to a directional blurring in the image. Its sources are usually clear, but occasionally it may be due to violent operation of the shutter or due to some one leaning heavily against the microscope during photographic exposure.

Instability of the High Voltage and Objective Lens Current Supplies

A change in HV supply manifests itself in a change in magnification, movement of the image, poor resolution, and a radial blurring.

A change in the objective lens current causes a rotational blurring of the image.

SELECTED BIBLIOGRAPHY

Agar, A.W.: The operation of the electron microscope. In Kay, D.H. (Ed.): *Techniques for Electron Microscopy*, 2nd ed. Philadelphia, F.A. Davis, 1965.

Agar, A.W., Alderson, R.H., and Chescoe, D.: *Principles and Practice of Electron Microscope Operation*. In Glauert, G. (Ed.): *Practical Methods in Electron Microscopy*, vol. 2. Amsterdam, North-Holland, 1974.

Chapman, S.K.: *Understanding and Optimising Electron Microscope Performance*. A Perkin-Elmer EM Publication.

Meek, G.A.: *Practical Electron Microscopy for Biologists*, 2nd ed. London, Wiley, 1976.

Sjöstrand, F.S.: *Electron Microscopy of Cells and Tissues*, Vol. 1. *Instrumentation and Techniques*. New York, Academic Press, 1967.

Wischnitzer, S.: *Introduction to Electron Microscopy*, 3rd ed. New York, Pergamon, 1981.

Chapter 8

PERFORMANCE TEST OF THE TRANSMISSION ELECTRON MICROSCOPE

THE ELECTRON MICROSCOPIST always wants to keep his instrument at the peak of its performance. Periodic instrumental evaluation that includes the determination of magnification and resolving power is, therefore, required at least once a month. The operator should keep a correct record of results of these tests for proper servicing of the instrument.

CALIBRATION OF THE MAGNIFICATION

The total optical magnification of electron images is the product of the magnifications of all the imaging lenses. Almost all modern instruments now have direct readout magnifications that are accurate to approximately ±5%. An automatic system for normalizing the magnification by standardized cycling of the lenses is available in some microscopes. The reproducibility of magnification in these instruments is better than ±2%. Without special precautions, however, the error can approach ±10%. The accurate determination of magnification of the transmission electron microscope (TEM) is considerably more difficult than the light microscope for the following reasons: (a) due to hysteresis, the field strength of electromagnetic lenses cannot be exactly calculated from the current flowing in them. The magnification of the image for given values of current in imaging lenses could, therefore, vary considerably; (b) the focal length of electron lenses varies with the fluctuation in accelerating potential and lens current. The focal length of the objective lens can vary due to different heights of the specimen holder; and (c) the exact position of the specimen plane may be different due to variations in the thickness of specimen, bent grids, and bent specimen holders.

The strength of the intermediate lens is usually varied to alter the magnification. Some microscopes have projector turret controls for selecting the proper projecter polepiece at the required magnification range. A specimen with known size is usually used for calibrating the magnification. The magnification can be calculated by the following formula:

$$\text{Magnification} = \frac{\text{Size of the image}}{\text{Actual size of the specimen}}$$

$$\text{Actual size of the specimen} = \frac{\text{Size of the image}}{\text{Overall magnification}}$$

The following methods are used for measuring the magnification of the TEM.

Polystyrene Latex Spheres

This is one of the oldest methods in which polystyrene latex spheres of very uniform size manufactured by the Dow Chemical Company were used for calibrating the magnification of the electron microscope. A large number of particles must be present across the plate for a reasonable accuracy. Latex particles may also be added to the specimen as reference particles for providing an internal standard for magnification and determination of size. However, the accuracy of calibration with these particles may vary more than ±10%, and this method is the least reliable. The particles of different batches vary in their diameter, and it has been almost impossible to produce particles of very uniform size. The latex particles can shrink under certain preparative conditions or after exposure to the beam.

Grating Replica Method

A replica made from a diffraction grating of known ruling spacing is most commonly used and is photographed at each instrumental magnification setting. Replicas of a cross-ruled diffraction grating or single-ruled grating are available, and a spacing of 2,000 lines/mm or 58,000 lines/inch is commonly used. A cross-ruled grating (Fig. 8-1) is preferable because distortion as well as magnification can be measured at the same time. As large a number of squares as possible should be measured, and eight to ten lines should be photographed on the plate for a reasonable accuracy. Calibration should be made in consecutive ascending order, starting with the zero setting (the diffraction spot) to minimize the hysteresis effect. The method is fairly accurate up to a magnification of about 50,000×, after which the crystal lattice spacing method should be used.

Crystal Lattice

Crystal lattice specimens are extremely useful for very high magnification calibration, and very accurate results can be reproduced. The lattice plane spacings of these specimens have been quite accurately determined by x-ray diffraction measurements. The measurement of magnification is similar to that of the replica grating method. Negatively stained beef liver catalase crystals (Fig. 8-2a) contain cross-lattice spacings of 87.5 and 68.5Å. Copper

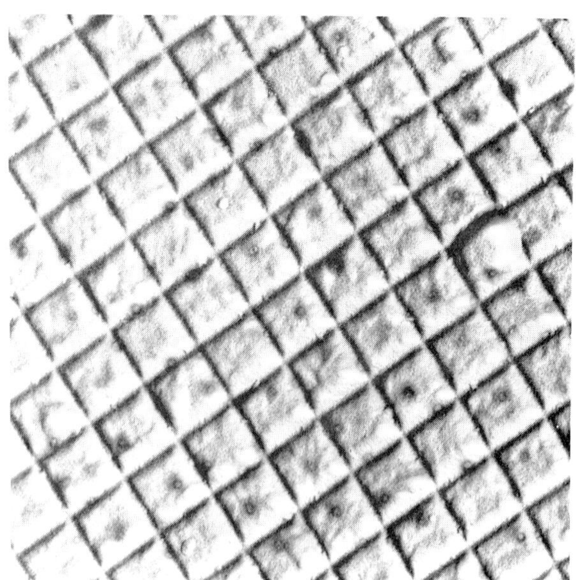

Figure 8-1. Magnification calibration with a replica of cross-ruled diffraction grating with 58,000 lines/inch. (×19,400)

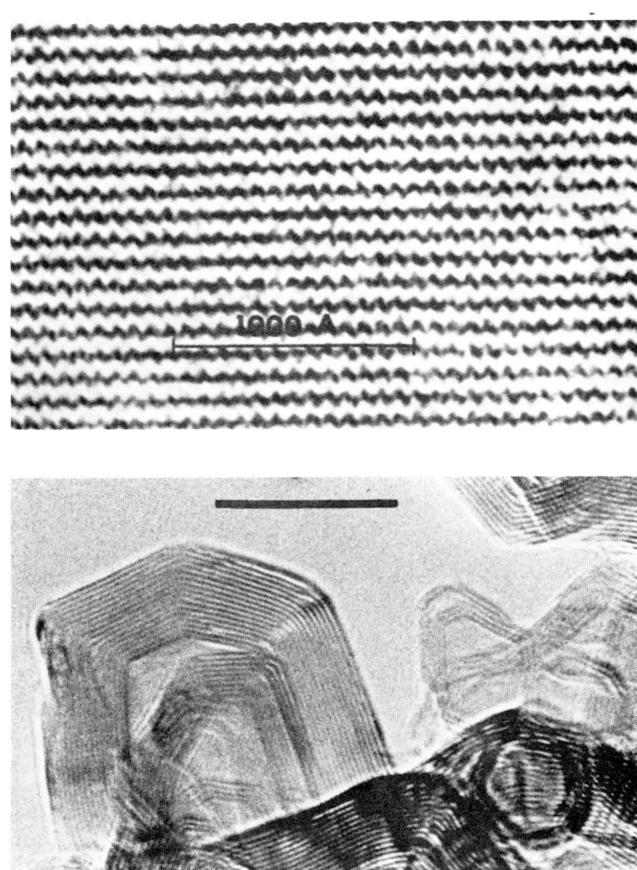

Figure 8-2. Micrographs of crystal lattice spacings. (a) Catalase. From R.W. Horne, The application of negative staining methods in quantitative electron microscopy. In G.F. Bahr and E.H. Zeitler (Eds.), *Quantitative Electron Microscopy*. Copyright © 1965. Courtesy of the Williams & Wilkins Co., Baltimore. (b) Graphite with lattice spacing of 3.4 Å. Bar = 100 Å. Courtesy of Siemens.

phthalocyanin with lattice spacing of 9.8 and 12.6 Å and graphite (Fig. 8-2b) with a highly stable lattice of 3.4 Å may be used for magnifications up to 500,000×. These specimens have to be examined very carefully with a low beam intensity because they are extremely susceptible to the beam, which causes the disappearance of the lattice.

DETERMINATION OF RESOLVING POWER

Practically, the *resolving power* is distinctly different from the *resolution.* Resolving power of a TEM is the property of the instrument that varies with the operational conditions and is limited by instrumental parameters. The resolution achieved in the image can be measured after photographing a suitable specimen. *Effective resolution* or *specimen resolution* of a biologic material, then, depends on the resolving power of the instrument, the specimen qualities, and the conditions of focusing. The resolution guaranteed by the manufacturer can seldom if ever be achieved with biologic specimens, and the definition of *high resolution* is somewhat controversial. Details of structures of 5Å or less are difficult to interpret because the fine structures are enhanced by Fresnel fringes when the objective lens is slightly out of focus. Slightly defocused electron images also show apparent granular structures (defocus granularity) at very high magnification that can be misinterpreted as resolution.

An objective measurement of resolution in transmission electron microscopy is quite difficult. It should be emphasized that a high-resolution work is not possible without a suitable specimen. The specimen must be sufficiently thin to transmit enough electrons, and the image must have a good contrast. To obtain the highest resolution, the operator should be proficient in operating the microscope as well as in preparing suitable specimens. The high-resolution work also demands a lot of patience from the microscopist.

The microscope must be in an optimal condition and at the peak of its performance before a high-resolution work is undertaken. The following points should be checked: (a) the microscope column must be clean, especially the condenser and objective apertures, and the specimen holder; (b) the vacuum must be high, and the anticontaminator (*see* in this Chap.) must be filled with liquid N_2 and allowed to settle for at least half an hour before the resolution determination or high-resolution work, otherwise a temperature gradient may cause specimen drift; (c) the specimen drift should be checked at the highest magnification; (d) the microscope should be perfectly aligned, and the second condenser and objective lenses must be deastigmated; and (e) the electronics must be stable, and the fluctuation in high voltage and objective lens current should be within the limits set by the manufacturer. Approximately an hour before the resolution test or high-resolution work, the microscope must be warmed up with the high voltage switched on but the filament current off. The following procedures are used for measuring the resolving power.

Point Separation (Point-to-Point) Method

In this method, suitable small particles such as platinum-iridium evaporated on a very thin carbon film are photographed at a very high magnification. The print is then searched for two particles that can be distinguished as separate. The distance between their centers is the resolution (Fig. 8-3a) achieved on the photograph and is calculated by the following formula:

$$\text{Resolution} = \frac{\text{Smallest distance between the center of two distinct particles}}{\text{Overall magnification}}$$

To insure a negligible effect of electron noise, the particles should be photographed twice with the same instrumental settings, and the same pair of particles must be present in both the plates; otherwise, background granular structure can be confused with particles. Several pairs of particles in different orientation should also be photographed to demonstrate that astigmatism and other image defects are insignificant. This method is the most generally acceptable measurement of resolution because the test specimen used in this technique is very similar to that used in routine work. It is apparently not very accurate for resolution better than 5Å.

Lattice Spacing Method

The lattice resolution test is also widely used, and the resolution is determined by imaging the spacings in a crystal lattice plane. Lattice spacings of gold (Fig. 8-3b) or graphite (Fig. 8-2b) are ideal for high-resolution test. The clearly defined lattice lines can be more easily measured than the point spacing method for an extremely accurate resolution determination. The lattice plane is imaged due to interference between the unscattered primary beam and an appropriate diffracted beam. The technique may be unreliable as a guide

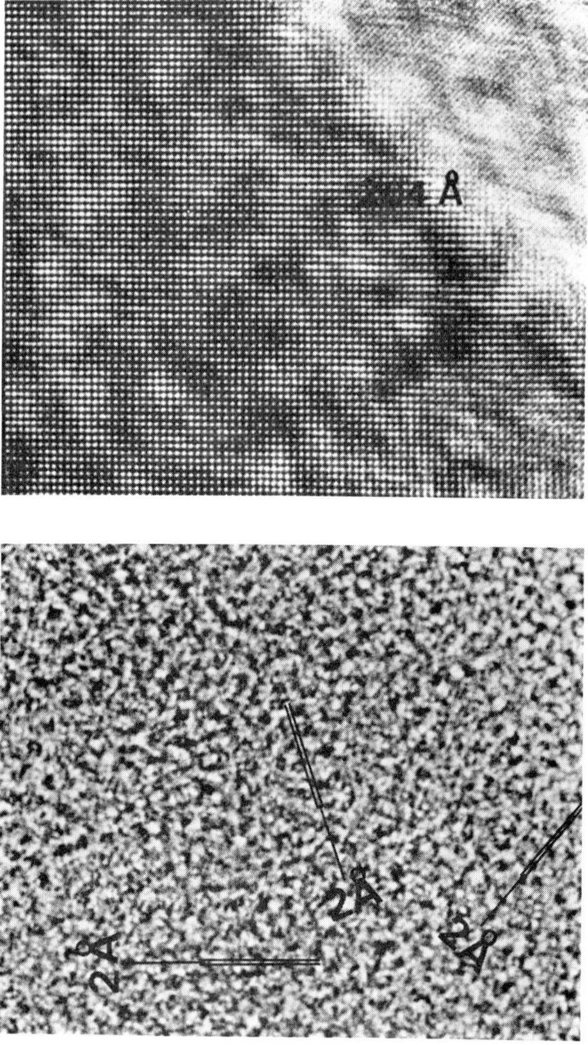

Figure 8-3. Resolution measurement. (a) Point-to-point (point separation) method with graphitized carbon showing a resolution of 2Å. (b) Gold crystals (200) lattice with 2.04Å spacings. Both courtesy of Hitachi, Ltd.

to the resolution obtained with conventional specimens. This is because the two-beam image-forming system is not affected by the spherical aberration to the same extent as in conventional electron microscopy. In actuality, the measurement of lattice spacing is not necessary because the technique required for imaging the lattice plane determines the lattice spacing resolved. It should be remembered that crystal lattice specimens are highly beam sensitive, and their examination requires experienced operators.

Fresnel Fringe Test

The minimum resolvable overfocus fringe around a hole in a carbon film is a fairly accurate criterion of the resolving power of the instrument (Fig. 8-4), and it is sensitive up to 4Å. A through-focal series of micrographs of a clean nearly circular hole in a holey carbon film is taken at a very high magnification. One of these plates should have the uniform overfocused fringe just touching the hole edge. It is carefully measured with a graticule magnifier at different

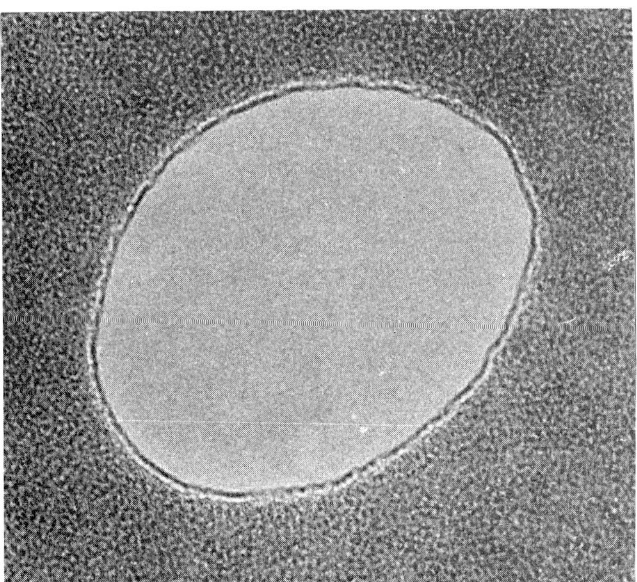

Figure 8-4. Resolution determination by the Fresnel fringe test. The total magnification of the micrograph = 600,000×. The average fringe spacing = 0.8 mm. Therefore, the resolution = $\frac{0.8 \text{ mm}}{600{,}000}$ = 13.3Å.

points, and the fringe width is averaged (maximum width + minimum width/2) to determine the resolution. The resolution can be determined by the following formula:

$$\text{Resolution} = \frac{\text{Average distance between the centers of the dark fringe and white fringe}}{\text{Overall magnification}}$$

The Fresnel fringe varies in width with variable parameters of the electron microscope, and as such they are also the most useful guide to the instrumental stability and its performance. The stability of the power supplies and electronics can be checked by watching the fringes close to focus for a few minutes. The movement of the fringe in and out from its mean position indicates an instability.

CONTAMINATION AND MEASUREMENT OF ITS RATE

Decomposed products of the vacuum system are deposited as *contamination* on the interior surfaces of the columns during the operation of the electron microscope. Contamination is most noticeable on the surface of the specimen. The major source of the contaminants is residual gases and water vapors in the vacuum system because the vacuum in the column is not absolutely clean. Air molecules are present even at 10^{-6} torr, the best vacuum achieved in a conventional TEM. Vapors are also introduced by photographic films. A large number of hydrocarbon molecules are derived from vacuum pump oils, grease from rubber *O* rings that are used as vacuum seals, and from the solvents used for cleaning the microscope components.

The hydrocarbons settle on the specimen, forming amorphous chalky deposits. After exposure to the electrons, the hydrocarbon molecules are broken down to carbon and hydrogen. Hydrogen reenters the column, but carbon remains on the specimen surface, and this *carbon contamination* (Fig. 8-5) can grow rapidly to as high as 5Å/second under favorable conditions. This reduces the resolution and contrast and obscures fine structures of the specimen. Furthermore, a contamination deposit on objective apertures and polepieces would act as electrostatic lenses and introduce astigmatism. Therefore, the interior components, especially the apertures and polepieces, would require periodic cleaning.

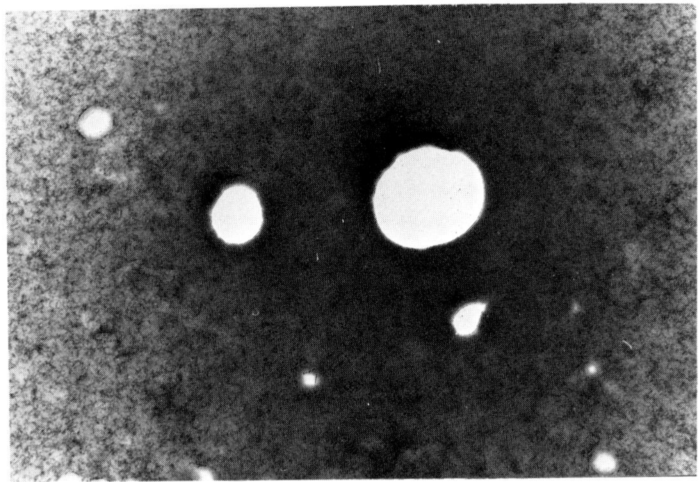

Figure 8-5. Contamination deposit on a thin carbon film. (×52,000)

The ionization of residual water vapor by electrons also produces hydroxyl ions. These ions react with the carbon in the biologic specimen, and a layer of specimen is vaporized away in the form of CO or CO_2. This is called *etching or stripping* (Fig. 8-6).

Reduction of Contamination

Contamination of the specimen can be reduced by reducing the intensity of the beam, which, of course, makes the image dim and difficult to focus. A cold trap or baffle at liquid N_2 temperature can be placed between the diffusion pump and column to prevent the hydrocarbon and water vapor from reaching the specimen surface. This, however, reduces the pumping speed. Another method is to heat the specimen above 200° C to drive off the contaminants, but this is not possible with biologic specimens.

By far the most successful method for reducing specimen contamination is to surround the specimen with a cold trap maintained at liquid N_2 temperature (−160° C to −180° C). A very drastic reduction in partial pressure of hydrocarbons near the specimen then occurs, and the cold surface preferentially attracts the contaminants to adsorb onto its surface. The modern high-performance instruments provide this device, called *cold finger* (Fig.

Figure 8-6. The effect of etching on a thin carbon film. Note etching around the hole. (×52,000)

8-7), either above or below the specimen, which is cooled by a Dewar flask filled with liquid N_2. In the side-entry stages, the specimen is sandwiched between two thin blades kept cold with liquid N_2. An efficient design of this type can reduce the contamination rate to less than 0.1Å/minute. A liquid N_2 reservoir in a metal cylinder making contact with the specimen chamber, provided in some older instruments, is not as efficient as the cold finger although it still reduces the contamination rate by a considerable factor.

Recently, *cryopumps* that surround the specimen have been introduced. They consist of larger surfaces cooled to liquid N_2 temperature and are used in the specimen chamber.

Measurement of Contamination Rate

The rate of contamination deposit on the specimen can be determined by using a holey carbon film. The instrument is set for normal operating conditions and for normal viewing intensity. A

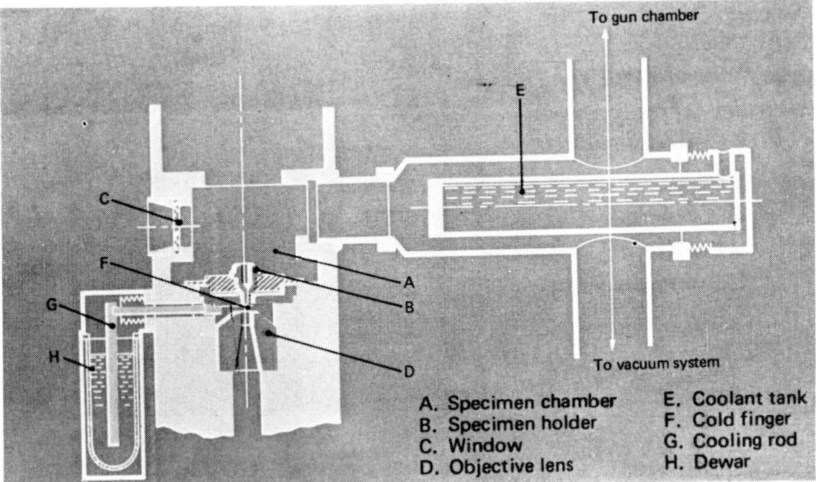

Figure 8-7. A schematic diagram of the anticontamination device (cold finger). Courtesy of Hitachi, Ltd.

nearly circular hole of 2 to 3 cm in diameter at an instrumental magnification (about 60,000×) is located, and a slightly underfocused image of the hole is photographed. The plate is left out, and the beam is allowed to illuminate the hole at the previous viewing intensity (which should have been previously noted by reading the intensity meter) for a specified period of time (usually 10 minutes). The microscope is not touched during this period. After this time, a second exposure of the hole on the same plate is made (superimposed image), and the plate is developed (Fig. 8-8). For accuracy, the rate should be measured over a period of about one-half hour, and several plates should be exposed. The contamination rate is calculated by the following formula:

$$\text{Contamination Rate} = \frac{D_1 - D_2 \times 10^8}{2 \times M \times t} \text{ Å/minute}$$

where D_1 = original diameter (in cm) of the hole; D_2 = its diameter (in cm) after the time t (filled-in hole); M = magnification, t = time in minutes.

The measurement of D_1 and D_2 is best made directly from the plate or the film. A graticule magnifier should be used if the hole is

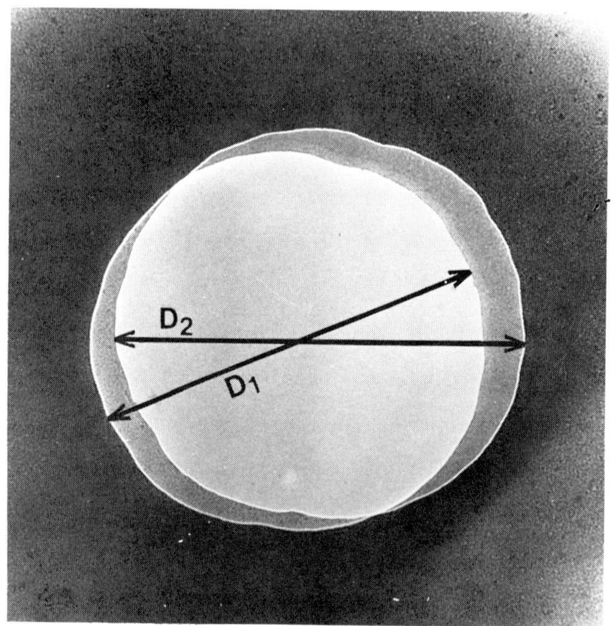

Figure 8-8. Measurement of contamination rate. The deposit of a layer of contamination is evident by the reduction of the diameter (D_2) of the superimposed hole; D_1 is the diameter of the original hole. The micrograph also shows a drift of the specimen during the ten-minute exposure. (×90,000)

small. Since a few holes are circular, D_1 and D_2 should be measured in two directions. The time interval required will depend on the contamination rate. If the rate is high, a period of a couple of minutes should be adequate, otherwise a ten-minute period may be needed. The rate should be measured with and without the anticontaminator and the results recorded. The rate is usually very high after dismantling and cleaning of the microscope column.

The contamination rate measurement test also indicates the specimen drift during the time it is exposed to the beam (Fig. 8-8), and the rate of specimen drift can also be calculated.

SELECTED BIBLIOGRAPHY

Agar, A.W., Alderson, R.H., and Chescoe, D.: *Principles and Practice of Electron Microscope Operation.* In Glauert, A. (Ed.): *Practical Methods in*

Electron Microscopy, vol. 2. Amsterdam, North-Holland, 1974.

Chapman, S.K.: *Understanding and Optimising Electron Microscope Performance*. A Perkin-Elmer Publication.

Horne, R.W.: The application of negative staining methods in quantitative electron microscopy. In Bahr, G.F. and Zeitler, E.H. (Eds.): *Quantitative Electron Microscopy*. Baltimore, Williams and Wilkins, 1965.

Meek, G.A.: *Practical Electron Microscopy for Biologists*, 2nd ed. London, Wiley, 1976.

Sjöstrand, F.S.: *Electron Microscopy of Cells and Tissues*. Vol. 1. *Instrumentation and Techniques*. New York, Academic Press, 1967.

PART II

TECHNIQUES

Chapter 9

PHOTOGRAPHY IN ELECTRON MICROSCOPY

THE RECORDING OF THE VISUAL image, the output of an electron microscope (EM), is of vital importance because the photographs of an object are the only permanent records of a specimen. The resolution of the image on the screen is poor due to the large grain size of the fluorescent coating, and the contrast of the reflecting screen is low. Furthermore, the specimen cannot be indefinitely examined in the EM because of the radiation damage and contamination. A photographic recording of the image provides a much higher resolution than the screen and is required for detailed evaluation, duplication, publication, and display.

A very expensive high-performance instrument along with the high cost for supporting the EM laboratory would be of little value if the operator cannot take good pictures, develop, and print them properly. Therefore, there is a saying that "the wedding between electron microscopy and photography is one in which no divorce is possible." Photography, like the EM itself, is a working tool of the microscopist through which he communicates his findings.

PRINCIPLES OF PHOTOGRAPHY

The photographic process consists of (a) exposing a suitable photographic emulsion (radiation-sensitive material) to image-forming electrons; (b) developing the exposed material to yield a negative, which is fixed, washed, and dried; and (c) printing the negative usually by enlargement onto photographic paper, which is then developed, fixed, washed, and dried in the same way as the original negative.

The photographic material consists of a glass plate or flexible film coated with a layer of emulsion, 20 to 40 μm thick. The emulsion is a layer of gelatin in which are embedded radiation-sensitive silver halide (silver bromide) crystals 100 to 500 nm in diameter. It is overcoated with a thin layer of hardened gelatin to minimize abrasion. Energy absorbed by the silver halide crystals from the incident radiation results in the formation of an invisible latent image. Subsequent chemical processes called *development* and *fixing* convert the exposed silver halide crystals to metallic silver, and a visible image of black metallic silver grains is formed. The optical density, D, of a photographic negative is

$$D = \log_{10} \frac{I_o}{I}$$

where I_o = the intensity of the incident radiation; I = the intensity of the transmitted radiation.

Exposure (E) can be defined as the darkening effect of the radiation on silver halide crystals and is expressed as follows:

$$E = I \times T$$

where I = the intensity of radiation and T = length of exposure.

Electron Image Recording

The process of latent image formation by electrons is significantly different from that of light. The radiant energy of photons of light that exposes the photographic material is equal to 2 to 3 electron volts, whereas the electron energy in the transmission EM (TEM) is 20 to 100 KV. The emulsion, therefore, responds quite differently to electrons than to photons of light. Each high-energy electron is capable of making ten or more silver halide grains developable. On the other hand, ten or more low-energy photons must be absorbed in each grain before it can be developed. When an emulsion is exposed to light, the density increases with the exposure up to a point, after which more exposure does not cause any more increase in density (the density shoulders off). However, with electrons, the density increases with the length of the exposure and does not shoulder off.

The *resolution* of the fine grain emulsion for electron microscopy is usually high to record fine details of the image. Some scattering occurs when the electrons enter the emulsion, and the electrons at 100 KV penetrate a greater distance in the emulsion before being scattered. This *spread function* is approximately 5 to 10 μm for most photographic emulsions used in electron microscopy, and the maximum enlargement that can be made is about 20×. The electrons must be slowed down before they can expose the emulsion, and to avoid a limited exposure, the emulsion must be thick.

By their very nature, electrons strike the emulsion at random and produce a fluctuation in density or *noise*. The image is then said to be *noisy* or *grainy*. The photographic emulsion can also introduce noise or *granularity*. The *detective quantum efficiency* (DQE) expresses the measure of the relative contribution of electron noise and emulsion noise as follows:

$$DQE = \frac{(\text{Signal/Noise})^2 \text{ in photographic record}}{(\text{Signal/Noise})^2 \text{ in the electron beam}}$$

The signal-to-noise ratio is a measure of the image quality of the electron beam, and the signal increases in proportion to the number of electrons incident on the specimen. The best way to improve the signal-to-noise ratio of the electron image, then, is to collect more electrons transmitted through the specimen. This can be done either by lengthening the time of exposure or by increasing the intensity of the beam. Longer exposure produces less noise or graininess, a denser negative, and more important, a higher contrast. A high-intensity beam also introduces less granularity, but the beam is usually less coherent and the image may not be sufficiently sharp.

The photographic material has also its own signal-to-noise ratio. The signal is the optical density, and the noise or granularity is the fluctuation in density from area to area. A DQE value of 1 indicates a perfect recording.

A photographic negative appears to consist of minute particles when viewed in an enlarger, which are called *grains*. While referring to these grains, *graininess* is used, which is mistakenly considered as a defect in the emulsion. Graininess, however, is not a defect; it is a record of the electron fluctuation or noise. This can be illustrated by exposing one plate to light and another to electrons. When the plates are developed to the same density, granularity is visible in the plate exposed to electrons but not in the plate exposed to light.

Contrast in photography refers to the tonal range of an image. In the electron micrograph, it is simply the difference in density between the darkest and lightest areas. Some enhancement of contrast can be achieved during the photographic process. This can be done by either increasing the exposure with no change in development or by increasing development without a change in the exposure.

Plates and Films

Traditionally, glass has been used as a base for photographic emulsion in electron microscopy. Glass plates are rigid, do not absorb water, dessicate rapidly, and often can be put into the TEM column without prior dessication. However, glass is brittle, bulky, and costlier than acetate-base roll films or estar thick base (polyester) cut films. On the other hand, films are easily stored, but they require a considerable time for dessicating them effectively; acetate films liberate a plasticizer, which contaminates the column, and their dimensional stability is somewhat poor. The roll films can be of 70 mm or 35 mm size. These films require higher optical enlargement than the plates or cut films. If scratches and other imperfections are present, they would become conspicuous in the final print.

Kodak electron image glass plates (size $3^1/_4'' \times 4''$), specifically designed for TEM image recording, have a very fine grain emulsion with a high DQE and are recommended for routine use in transmission electron microscopy. Kodak estar thick base cut film 4489 (same size as plates) is also quite satisfactory for this purpose and has become very popular among electron microscopists. Kodalith estar base LR film 2572 can be used for TEM image recording. Agfa-Gevaert, Fuji, and Ilford plates and films are also available for TEM work. The very fine grain blue sensitive photographic emulsions for transmission electron microscopy are less sensitive to light and can be handled under fairly bright yellow or red safe light.

Roll films are usually placed close to the projector lens, and the plates or cut films are placed under the fluorescent screen. A plate or film reserve chamber is usually provided in the microscope. This chamber is also provided with dessication facilities for roll films. These films can also be dried in a dessicator before being loaded in the column.

Exposure

The exposure time depends on the stability of the specimen. Ideally, it should be as short as possible to avoid blurring due to specimen drift and instrumental instability. A short exposure time, however, introduces graininess and may cause uneven exposure when the shutter is raised or lowered. This is not a problem with an electric shutter, which allows automatic exposure. An exposure time of 2 to 5 seconds is preferred because long exposures are less grainy and produce good contrast. Intensity of the electron beam and exposure time both build contrast into the negative. A very sensitive emulsion is required for a very short exposure if the specimen is sensitive to the beam. Contrasty specimens usually need longer exposure and shorter development. The image contrast is reduced as the magnification increases, and a shorter exposure with longer development is usually used for high magnification photography.

The electron image on the fluorescent screen is not generally photographed through the viewing window because of the limited resolution and contrast on the screen. A camera attached to the viewing window would interfere with normal image viewing. The screen also has to be tilted for photography, resulting in a distorted image. The image in the TEM is usually recorded by lifting the screen so that the electron image falls on the plate directly.

The procedure for exposing the plate has been described in Chapter 7. All modern instruments are equipped with an exposure meter, and all high-performance instruments have a shutter electrically coupled to the exposure meter and the correct exposure is given automatically. A lower intensity should be used to obtain a coherent beam. The intensity used for focusing has to be turned down to the desired photographic intensity. The exposure should be made firmly, and there should be no mechanical vibration during this time. In the absence of an automatic numbering system in the microscope, the plates should be numbered either with a numbering device or with a diamond-tip pen before being loaded into the cassettes. The emulsion side should face the electron source in the microscope, and the room lights should be turned off during exposure. To prevent double exposure, the plate should be drawn into the cassette receiver immediately after it is exposed. Full particulars concerning the micrograph, including the name of the operator, should be recorded after the plate is exposed (*see* Chap. 7).

Development

It is desirable to develop the plates or films in a room adjacent to the microscope room. They should not be developed in the microscope room because dust from dried-up spilled solutions can contaminate the column. The specimen should be left in the microscope while the plates are being developed so that it can be rephotographed if a satisfactory negative is not obtained.

The developer is an alkaline solution (usually pH 10 to 11) that reduces the exposed silver halide crystals to metallic silver. It usually contains metol, hydroquinone, sodium sulphite, sodium carbonate, and potassium bromide. The developer also slowly attacks unexposed silver halide grains that give rise to *background fog*. The developing solution usually contains reagents to minimize fogging and oxidation. It is preferable to use a fresh developing solution each time, but a developer stored in covered stainless steel or plastic tanks is usually active up to seven to ten days. As the time of development increases, the density of the negative also increases. Development can also be increased by the concentration of the solution and by its temperature. A higher temperature causes a quicker development, but it also causes some swelling of the gelatin that may produce *reticulation*; the gelatin may shrink in cold water, which causes *frilling*. The temperature of the developing solution should be 68° F (20° C).

Developing usually takes several minutes, and it is important to agitate the solution during this time so that the solution is frequently moved over the emulsion. This is easily carried out by lifting the racks out of the developer and replacing them back at 30-second intervals. Generally, the development time cannot be standardized, and correct development warrants the judgement and experience of the microscopist. The negative must have adequate density for contrast, but it should not be too dark. It should be remembered that *a plate can be developed only once and that must be correct.* A good print can be made only from a good negative, and no amount of doctoring can help a poor plate. If a negative is poor, it should be discarded, and a photograph should be retaken.

Development is best done in stainless steel or plastic tanks. The concentration of the solution should be adjusted so that development occurs in two to five minutes. Plates can be placed in individual stainless steel racks. Racks made of Plexiglas for developing eight to twelve plates or sheet films are also available. The roll film is usually wound onto a plastic wheel and processed in a small tank. Kodak D-19 developer can be used for good speed and minimum grain and is recommended for Kodak electron image plates and estar base cut films. Kodak recommends adding its antifog No. 1 to this solution. Kodak HRP and Ilford PQ universal developer are also available. Kodak dektol developer is used for roll films.

Fixing

After development, the unexposed silver halide grains must be removed from the plate to render the unexposed areas transparent. Otherwise, they would darken on exposure to light and obscure the image. This is done by a further chemical process called *fixing*. The *fixer* or *hypo* is an acidic solution that generally contains sodium or ammonium thiosulphate and potassium metabisulphite. It forms a soluble compound with silver bromide but has no effect on the silver image; it also hardens the gelatin. Kodak fixer can be purchased as a concentrated solution or as powder. Other rapid fixers are also available. An acetic acid stop bath or plain running water should be used between developing and fixing. This is done to minimize contamination of the fixer by the developer, to prolong the life of the fixer, and to insure precise termination of development.

The acid fixer clears the plates and hardens the gelatin in 3 to 5 minutes. After fixing, the chemicals and their byproducts must be washed off the negative in running water at 68° F (20° C) for 10 to 20 minutes. Otherwise, the silver image would fade with time. A wetting agent should be added to the final rinse (preferably distilled water) before drying. Kodak photoflo is good for this purpose, but any household clear liquid detergent may also be satisfactory. The plates can be conveniently dried in the rack. The negatives should be examined with a 5× hand lens on a lighted viewing box. They should be stored in transparent envelopes and filed.

Printing the Electron Micrographs

Ideally, there should be two darkrooms, one for developing the plates and the other for printing. A positive print of various sizes is made from the negative on photographic printing papers. Contact prints are rarely useful in electron microscopy, and it is essential to have an enlarger of very high quality such as Durst enlarger. The enlarger should have a double condenser system and a point-light source bulb preferably of the halogen-quartz type for high contrast. The opaque large white bulb has no place in the EM printing room because it reduces image contrast by scattering the light in the emulsion. The point-light source, however, shows any blemishes and imperfections that may be present in the negative, and the lens has to be fully open to compensate for this (aperture stops cannot be used). The intensity can then be controlled with a powerstat.

The image in the negative is made of discrete grains. The grain size varies with the density, and it may be fairly large in overdeveloped plates. Although Kodak electron image plates or estar base films can be enlarged up to 20×, the usual enlargement used by electron microscopists is usually 3 to 5×. The 35 mm films require an enlargement of at least 8 to 10×.

The plate or film is placed in the carrier of the enlarger with the emulsion side down. The image should be examined on the easel against a white background (the back of a 8″ × 10″ sheet of printing paper is often quite suitable). The image can be focused with the aid of a focusing target usually built onto the negative carrier. However, the most accurate focusing is achieved by focusing on the grain of the negative with a scoponet enlarger magnifier.

The choice of the right grade of paper is very important in printing. Various grades of Kodabromide single-weight papers ranging from No. 1 (soft) to No. 5 (very hard) are used. Multigrade Kodak polycontrast and rapid polycontrast papers are also quite satisfactory for this purpose, and the required contrast can be obtained by using the Kodak polycontrast filters. The advantage of this paper is that only one grade of paper is needed for all printing work. Different grades and multigrade Agfa-Gevaert, Fuji, and Ilford papers are also available.

Orange photographic safelights should be used in the darkroom, and the glossy side of the paper should face the light source in the enlarger. Electronic devices to determine the appropriate grade of paper and for exposing it automatically by measuring the contrast and intensity are available. High-contrast images are best printed on soft to medium grade paper. Low-contrast images, on the other hand, require hard paper for enhancing their contrast. The intensity should be adjusted to the proper level, and automatic exposure should be done with a timer. There should be no vibration at the base of the enlarger during the exposure. The exact intensity level and exposure time are determined by experience. *Dodging* or *shading* is sometimes used to compensate an uneven illumination in the negative and to improve the print. Electronic dodgers that are fairly expensive are available. This can also be done by waving the hand or a cut out card over very light areas during exposure.

Prints are developed, fixed, washed, and dried in much the same manner as the negatives. The exact time of development cannot be standardized, and the electron microscopist learns this through experience. The prints are best developed in a tray, and Kodak D-19, Dektol, D-72, or Ektaflo type 1 is used for developing the prints in the U.S.A. A print should not be taken out of the developer before it is fully developed although it may appear too dark in the orange safelight. The prints should then be placed in a dilute acetic acid stop bath (a must for papers) and then transferred to the acid fixer.

After fixing for 4 to 5 minutes, the prints should be thoroughly washed in running water (temperature 68° F) for at least 30 minutes because the paper base is soaked with chemicals. The prints are soaked in a solution of print flattener (Kodak print flattener or Pakosol) and then dried, preferably in a roller-drum type of drier. The drum should always be kept clean, and the glossy side should be toward the drum. The negative number and the times of enlargement should be penciled in on the back of the paper before the exposure is made. A wax pencil may also be used to mark on a corner of the glossy side that can be wiped off with a soft cloth without damaging the print. An efficient filing system should be used for the prints.

Automatic print processors such as Kodak autoprocessor are available that can produce dry prints rapidly (in 1 to 2 minutes) after the paper has been removed from the enlarger. They use a *stabilizer processing technique* in which two solutions, activator and stabilizer, replace the conventional developer and fixer. A special printing paper is used for this processing. There is, of course, no margin for error because the processing is fixed. The prints for these machines are good for initial evaluation, but they are not permanent. They fade with time and tend to curl because of the thick emulsion used. The prints can be made permanent by fixing, washing, and drying, but then the conveniences and advantages of the system are lost. The quality of these processors has been improved since they were first introduced in late 1960s.

SELECTED BIBLIOGRAPHY

Agar, A.W., Alderson, R.H., and Chiscoe, D.: *Principles and Practice of Electron Microscope Operation.* In Glauert, A. (Ed.): *Practical Methods in Electron Microscopy*, vol. 2. Amsterdam, North-Holland, 1974.

Chapman, S.K.: *Understanding and Optimising Electron Microscope Performance.* A Perkin-Elmer EM Publication.

Electron Microscopy and Photography, A Kodak data book. Eastman Kodak Co. 1973.

Meek, G.A.: *Practical Electron Microscopy for Biologists*, 2nd ed. London, Wiley, 1976.

Sjöstrand, F.S.: *Electron Microscopy of Cells and Tissues*, Vol. 1, *Instrumentation and Techniques.* New York, Academic Press, 1967.

Chapter 10

SPECIMEN SUPPORT FILMS AND MOUNTING OF SPECIMENS

THE PENETRATING POWER of electrons in the 20 to 125 KV range is low; therefore, very thin films, no more than 200Å thick, are used for mounting the specimens for electron microscopy. A very thin support film or substrate is the transmission electron microscope (TEM) analogue of the glass slide used in light microscopy. An ideal specimen support should meet the following criteria: (a) it must be highly electron transparent; (b) it must conduct heat from the specimen; (c) it should have no surface structures; (d) it must be mechanically stable under electron bombardment; (e) it must form a thin and flat surface; and (f) it must be wettable for spreading of the specimen.

PREPARATION OF THIN FILMS AND COATING OF GRIDS

Support films can be made from a variety of materials, but the most widely used films are plastics or evaporated carbon.

Specimen Grids

Grids are basic supports for thin films, and a wide variety of grid design is commercially available (Fig. 10-1). Originally, specimen carriers (apertures) made of thick platinum disks with multiple small holes that had to be cleaned after each use were employed. Later, grids punched out of fine wire mesh copper or brass sheets with 200 holes/linear inch were introduced. They were soon replaced by perforated electrodeposited grids that are inexpensive enough to be discarded after one use. Grids are generally made of copper and

Figure 10-1. Specimen grid designs (3.05 mm dia.) commonly used in biologic electron microscopy. The grids in the top row have 200 mesh/inch to 400 mesh/inch. The middle row consists of slotted grids, and the two grids in the bottom are represented by a grid with a single circular hole and a folding grid (100/300 mesh).

supplied in two sizes, 2.3 and 3.05 mm in diameter. For special needs, e.g. for treatment of specimens with highly reactive reagents and high temperature metallurgic work, stainless steel, platinum, nickel, gold, silver, palladium, tungsten, molybdenum, or rhodium grids are available.

Grids with square holes are available in meshes ranging from 50 to 500 holes/linear inch. The open area decreases as the mesh/inch increases. Grids with large open areas weaken the grid and cause specimen heating and charging. The 200-mesh grid is most commonly used, but a 300-mesh grid preferably with round holes is necessary for high-resolution work and specimen stability. Grids with central marks are convenient for orientation in the microscope. Grids are very thin, and they must be handled with very fine watchmaker's forceps (No. 4 or 5). They should be held at the edge, and picking up grids from a flat surface requires practice. The forceps tips should be sharpened with a fine emery paper to give a positive grip.

Plastic Films

Collodion (parlodion, nitrocellulose, Zapon) films have been used as early as 1936. It is easily cast on water and spreads very well, but it is not very stable under the electron beam, especially at a higher KV, and it is no longer popular. Formvar (polyvinylformal, pioloform F, Rhovinal F), the most popular plastic film, is stronger than collodion, but it does not spread well on water, and a very thin film is somewhat difficult to prepare. Films of both these plastics cannot be made less than 200Å thick, they move slightly when first irradiated, and they are not suitable for high-resolution work. A resin concentration of 0.2 to 0.5% is usually used, and the concentration determines the thickness of the film.

Collodion Cast on Water

A 0.5 to 1% (w/v) collodion, dissolved in amyl or butyl acetate, is generally used for making this film. A shallow dish (≈20 cm in dia.) is filled with water, and the surface of the subphase is cleaned of dust or dirt particles with a clean glass rod. Alternatively, a thick film of collodion can be made by depositing a few drops of collodion on water and allowing it to spread out and dry. The film is removed by a needle, and dust and dirt particles are picked up with the film. Two drops of collodion are now placed on water with a capillary pipette, and the film is allowed to dry. A somewhat thicker film with uneven thickness is produced by this method. It is possible to see the interference color of film with a reflected light. It should be dark gray; if it is yellowish or bluish, it is too thick and should not be used.

The easiest method of coating the grids with this film is to place a number of grids on top of the floating film. Grids may be cleaned with chloroform or alcohol before coating. Grids have a shiny and a matt surface, and some microscopists prefer to put the film on the matt surface for better adhesion. Proper adhesion is attained by pushing the grids onto the film with a forceps. Bent grids should not be coated. A number of methods are used for picking up the filmed grids.

The simplest method of picking up coated grids is to place a piece of filter paper (or any absorbent paper) over them. After the paper is completely wet, it is carefully lifted up, carrying both the film and the grid with it (Fig. 10-2). The filter paper can now be placed in a

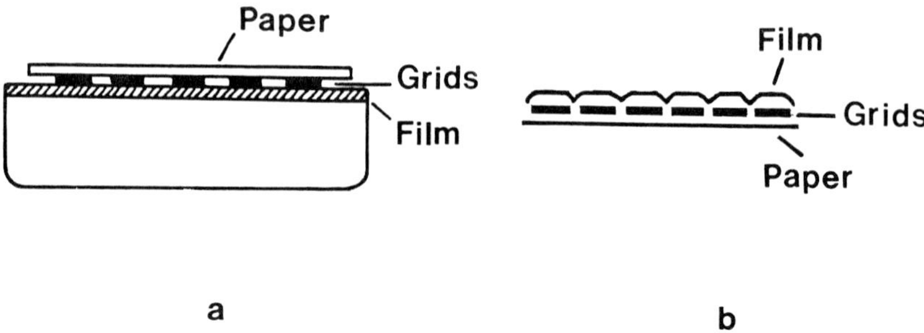

Figure 10-2. Preparation of filmed grids by placing filter paper on them. (a) Filter paper placed on the coated grids. (b) After complete wetting, the paper with adhering grids has been lifted from water.

covered petri dish (to protect from dust and dirt) until the film is dry. The grids can be dried at 40° C to shorten the drying time. After drying, the grids can be picked up with a forceps and stored in a variety of grid-storage containers or left adhering to the paper for convenience in subsequent handling.

Alternatively, a glass slide can be brought down vertically over the film. As the slide is pushed down past the surface, the film folds back against the slide. The slide is then turned and lifted off the water; excess water is blotted by touching the back and edge to filter paper and allowed to dry.

A *scooping technique* may also be used for the purpose. In this method, a scoop covered with a fine wire mesh or a grid spade is lowered into the water, and as many as 30 to 50 grids can be placed on it. The scoop is slowly lifted upward, bringing it against the underside of the floating film, and then lifted clear of the water surface (Fig. 10-3). Excess water is blotted off, and the grids are allowed to dry. An arrangement can also be made for drawing the water from the bottom of the dish so that the film settles down over the grids on the scoop.

Another method is to roll up the filmed grids on a clean glass bottle. The bottle is placed across one end of the film, rotated, and lifted clear of water.

Formvar Cast on Glass Slide

Although formvar films are stronger than collodion, they cannot

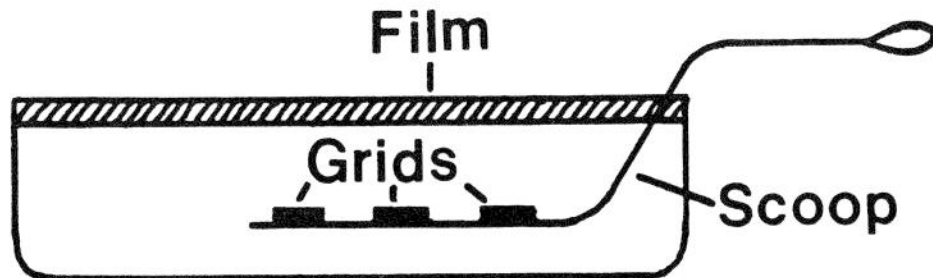

Figure 10-3. Preparation of filmed grids by the scooping technique.

be prepared by casting on water because, unlike collodion, formvar does not spread well. A *stripping technique* is used for preparing this film. A 0.2 to 0.5% (w/v) formvar solution in ethylene dichloride or chloroform is used. A precleaned light microscope glass slide is dipped into formvar kept in a coplin jar or a small beaker. Excess formvar is drained off by touching the edge of the slide to a filter paper, and the slide is allowed to air dry thoroughly. When the film is completely dry, it is scratched at the edge with a needle and floated on water by slowly lowering the slide into water at a shallow angle. The surface tension force strips the film off the glass. Breathing on the film prior to immersion in water may facilitate release. The film is mounted on grids by any one of the methods described for collodion.

Thick films do not adhere properly; they affect the resolution and cause specimen drift. Very thin films are unstable in the beam, and they must be strengthened by evaporating a thin layer of carbon onto them. Formvar-coated grids on the filter paper, glass slide, or the scoop are carbonized by simply placing them in a vacuum evaporator (Fig. 10-4) and by evaporating a thin layer of carbon on top of the formvar. These formvar-carbon films are most popular among biologists.

Evaporated Carbon Films

Plain carbon films introduced by Bradley in 1954 are best for high-resolution work. They are strong and stable in the microscope, their thickness can be controlled, and they conduct heat and electricity and thus reduce specimen drift. Unlike plastic films, which are simple to prepare, however, preparation of carbon film is tedious, time-consuming, and at times irreproducible.

A plain carbon film is prepared by vacuum evaporation of carbon. A precleaned glass slide or a freshly cleaved sheet of mica is

Figure 10-4. A typical vacuum evaporator for carbon coating and shadowing.

used; carbon films can also be prepared on glycerol. The evaporating apparatus consists of two pointed spectrographic carbon arc rods mounted into insulated holders (Fig. 10-5). One rod is usually fixed and the other movable; the points of these rods press slightly against each other by means of a spring. The end of one rod may be sharpened and the end of the other rod may be flat, or both rods may be ground. The target is placed 10 to 15 cm away from the source. A simple thickness indicator consisting of a piece of glazed porcelain with a drop of vacuum oil is placed near the target. The bell jar is evacuated (10^{-4} to 10^{-5} torr), and a 50A current is passed through the points for about half a second. Many evaporators have a pulsing device for this purpose. A just-detectable brown color on the portion of porcelain not covered with oil indicates a carbon coating of approximately 50A thick.

The film on the glass slide or mica is then scored with a needle and floated onto water at a shallow angle. The film can be scored into small squares, and breathing on the film prior to immersion may facilitate stripping. Carbon film is very brittle, and a small piece of

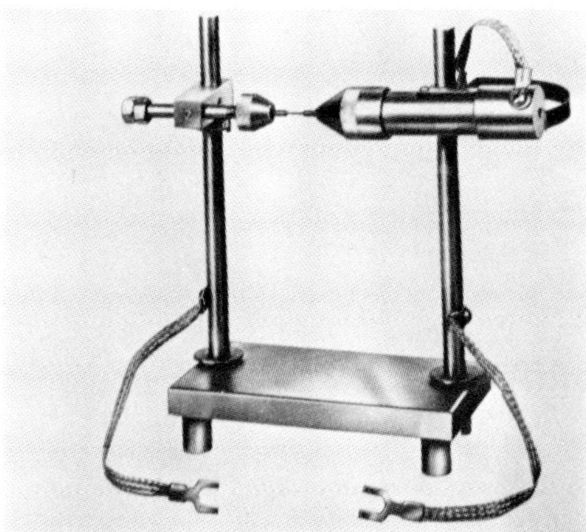

Figure 10-5. A carbon rod evaporation unit.

the film has to be picked up by immersing a grid held in a forceps (*fishing technique*). Grids immersed in concentrated HCl for a few seconds and then washed in distilled water facilitate proper adhesion of the carbon films, probably due to a slight etching caused by the acid. Before drying the grids, any pieces of the film that are folded under them must be removed with a needle. A number of grids can also be coated by the scooping technique previously described.

Carbon films are generally hydrophobic and not wettable, but upon aging they become hydrophilic. A low vacuum pressure and contaminating oil vapors during evaporation apparently make these films more hydrophobic. The contaminating layer can be removed and the film made wettable by touching the coated grids to an organic solvent, e.g. chloroform, or by a glow discharge in air at low pressure.

For very high resolution work, some electron microscopists prefer very thin carbon film made by dissolving the formvar backing from the formvar-carbon film. This can be accomplished by the following procedures. A small petri dish is filled with ethylene dichloride or chloroform. The formvar-carbon coated grid is held in a forceps, film side up, and brought to touch the surface of the solvent. After a few seconds, the grid is removed and allowed to dry while still being held in the forceps. After repeating the process, the grid is fully immersed in the solvent, and then the film would not float off. The grid should be immersed for 15 minutes to dissolve the plastic completely. It is then removed and dried.

The carbon film can also be freed from the plastic backing by the *flow-wash method.* The grid is slightly bent to a convex shape and placed on a metal peg. A burette is filled with the solvent and is placed 1 mm above the grid. A drop of acetone is placed on the peg, and before the acetone dries, the nozzle of the burette is opened to allow the solvent to flow over the grid at the rate of approximately 1 ml/minute. After 1 minute, the peg is removed and the grid allowed to dry.

Other Evaporated Films

Silica or silicon monoxide films can also be made by evaporation, but they are difficult to see and are more brittle than carbon. Aluminum oxide films can also be made, but they have background structures. Films made by evaporating single crystals of graphite have been used for extremely high resolution work.

Preparation of Holey (Perforated) Films

These films are used for correcting the astigmatism, determining the resolution, focusing, and for determining the stability of the electron microscope. Formvar-carbon holey films (Fig. 10-6) with nearly circular holes ranging from 0.1 to 1 μm in diameter are most suitable. The plastic backing can be dissolved for obtaining holey carbon films. A formvar net with holes ranging from 1 to 10 μm are used as a support for very delicate specimens such as crystal lattices and ultrathin sections. Carbonized formvar nets are ideal supports for thin sections.

Many methods for preparing holey films have been introduced, but the preparation of these films is often unreliable, chancy, and irreproducible. Although holey collodion films can also be made, holey formvar films are most commonly used. Basically, water droplets that pierce the film are deposited on a wet formvar film, making the holes. The water content affects the hole size.

In the simplest method, a clean glass slide is dipped in a 0.2 to 0.3% formvar solution in ethylene dichloride, and water droplets are immediately deposited by breathing on the wet film. The film becomes milky due to condensation of water vapors. The film is

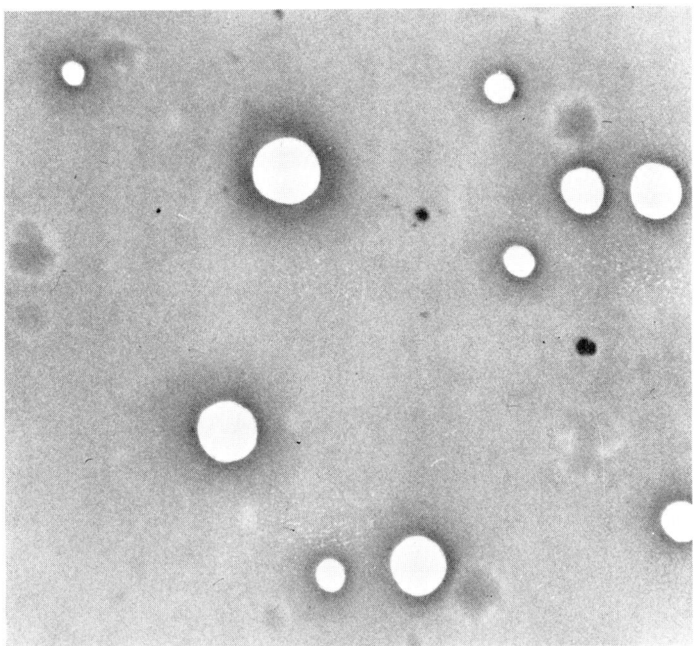

Figure 10-6. A formvar-carbon holey film with holes that have clearly demarcated edges. (×9,000)

allowed to dry thoroughly; it is then floated on water (by the stripping technique) and picked up on grids. The formvar concentration must be adjusted to make a very thin film, otherwise the water droplets would not cut through the plastic completely, and pseudoholes (Fig. 10-7) would be produced that are useless. The grids should be examined at a low magnification for determining the suitability of the film. It is important that the holes not run into one another and their edges be smooth.

For the second method, glycerol is added at a concentration of 1:8 or 1:16 to 0.25% formvar in ethylene dichloride or chloroform. The mixture is shaken, a glass slide dipped, and allowed to dry. The slide is held over boiling water or exposed to a jet of steam for 1 minute. After drying, the film is floated on water and collected on grids. The grids may have to be dipped in acetone for a few seconds to avoid the production of pseudoholes.

Alternatively, 1 to 5% water is added to a mixture of dioxane and

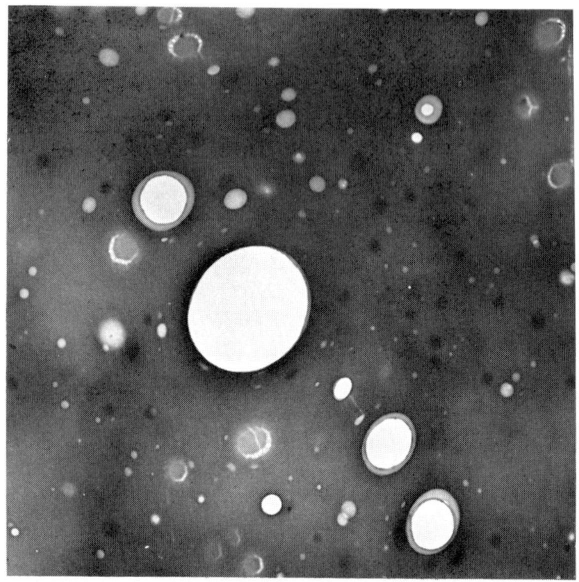

Figure 10-7. A holey carbon film with pseudoholes. (×9,000)

1% formvar in ethylene dichloride (1:1). The mixture is shaken, a film is made by casting on a glass slide, and the grids are coated. Dipping in acetone for a few seconds may be necessary for producing completely pierced holes.

A more reproducible method is to place a slide dipped in 0.2 to 0.4% formvar in a humidified petri dish (relative humidity 60 to 70%) at room temperature. A filter paper soaked in water is placed in the petri dish, and the slide is placed on the filter paper supported by two applicator sticks. Depending on the size of the holes desired, the slide may be kept from 5 to 10 minutes. After this time, the film on the slide is removed and allowed to dry thoroughly. The film is then floated on water, and the coated grids are immersed in acetone, absolute methanol, or chloroform-absolute methanol (1:9). The grids are kept in the solvent for 15 minutes to 1 hour at room temperature. They are then washed in 95% ethanol, 50% ethanol, and double distilled water (about 30 dips each). The edges of many holes formed at 60 to 70% relative humidity may be irregular, and some pseudoholes may be present. A mixture of chloroform-methanol is preferable, and the solvent smoothes the irregular edges and enlarges perforations. The chloroform ratio may be increased if adequate dissolution is not obtained.

A holey formvar film should be carbonized for stability. A formvar net (Fig. 10-8) can be made by exposing the wet formvar slide to water vapors for a longer time.

MOUNTING THE SPECIMEN ON COATED GRIDS

Biologic specimens generally consist of suspensions, thin sections, or replicas. Submicroscopic particles such as viruses suspended in a liquid have a tendency to aggregate when placed on the support film. The specimen must be dry, and any liquid from the specimen must be blotted off, otherwise water would scatter the beam and cause poor contrast. The specimen should also be stable in vacuum and the electron beam. Although volatile liquids that leave no residues, e.g. 1 to 2% ammonium acetate, ammonium carbonate, ammonium benzoate, have been used for suspending particles, triple sterile distilled water is one of the best all-around vehicles for this purpose.

Viruses generally require prior concentration and purification for examination in the TEM. They can be concentrated by ultracentri-

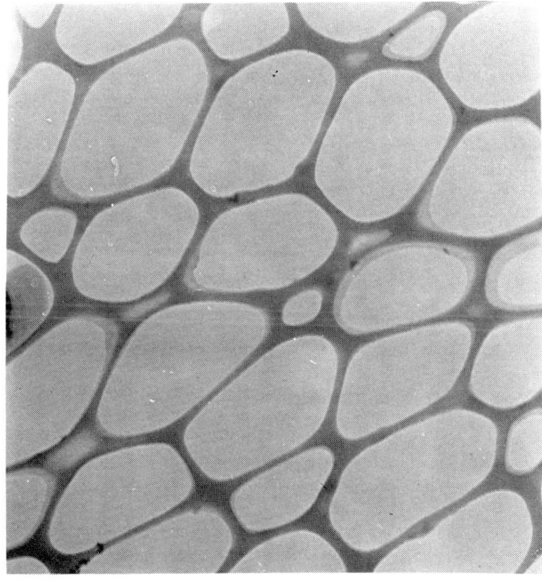

Figure 10-8. A formvar net. (×6,500)

fugation, and although a partial purification may be obtained during this procedure, a density gradient purification and dialysis are recommended for best results. Growing bacterial or protozoan cultures have to be washed for removing the growth medium. The wettability of liquid suspensions for better spreading on the film may be increased by adding bovine serum albumin (0.005 to 0.05%). Bacitracin at a concentration of 50 to 200 g/μl provides very good spreading, especially for negative staining (*see* Chap. 12). A high particle concentration is required for finding a fair number of particles in a viewing field, and a concentration of 10^8 particles/ml is recommended. All biologic specimens also need some form of preparation for increasing their contrast and for a better resolution (*see* Chap. 11).

The simplest procedure for depositing an aqueous specimen suspension on coated grids is as follows. A grid is held with a clamped forceps or with a forceps with an *O* ring (Fig. 10-9). A small drop of suspension from a Pasteur pipette is applied on the grid. It is better to prepare very fine capillary pipettes by drawing a glass tubing on a Bunsen flame. The drop is allowed to settle for 1 to 2 minutes, and the liquid is removed by touching with a filter paper. A drop of liquid usually remains behind the points of the forceps, even if the grid is dry. Therefore, when the grid is placed on a surface, the liquid causes it to move back up the points of the forceps, which results in the loss of the specimen. Before releasing the grid, the drop

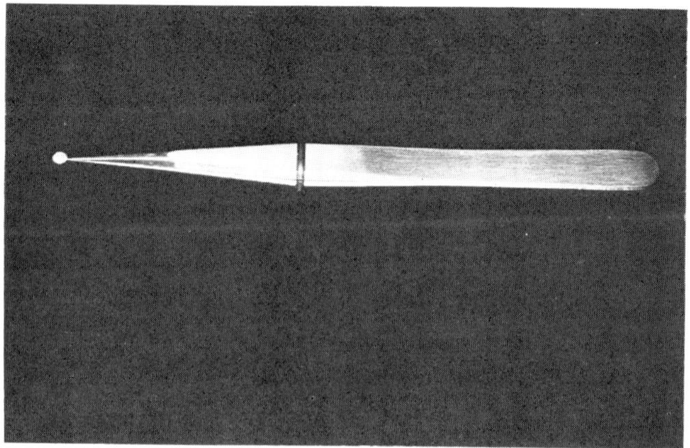

Figure 10-9. A clamped forceps holding a coated grid for mounting the specimen.

should always be removed with a filter paper. Sometimes it is necessary to fasten the grids to a glass slide. A narrow strip of cellophane tape is applied on the slide, and the grids are pushed under the tape so that they are held by the tape at their edges. A double-stick tape may also be used instead. They would be held tight during subsequent handling, e.g. for shadowing (*see* Chap. 11).

A welded platinum loop may also be used to place the liquid suspension on grids. A loopful of the suspension is applied on the grid, the liquid is allowed to settle for 1 to 2 minutes, and the fluid is then blotted off. The disadvantage of the drop method is that due to the large surface tension, the particles may become distorted or disassociated during the final stage of drying, and poor dispersion may occur.

Alternatively, the aqueous suspension can be sprayed onto a coated grid with a glass atomizer. A vaponephrin nebulizer attached to a funnel (Fig. 10-10) is quite suitable for this purpose. This is one of the most effective methods for dispersing a specimen suspension. The spray droplets dry instantaneously, and the specimen is ready to be examined. A high-velocity spray gun may also be used. A concentration of 10^{10} particles/ml is needed for atomizing. Special precautions should be taken for spraying infectious materials for which the spraying system should be a closed type and the parts should be suitable for decontamination.

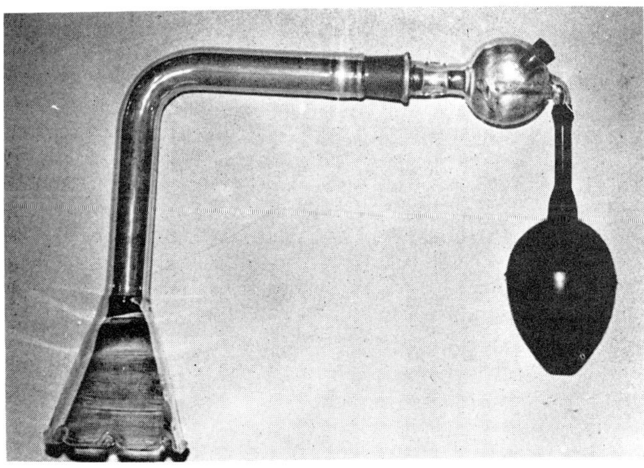

Figure 10-10. The vaponephrin nebulizer unit for spraying the specimen on grids.

Ultrathin sections can be placed on uncoated (bare) grids or on grids coated with a formvar-carbon net. Replicas are collected on bare grids while floating on water (*see* Chap. 11). Powdered materials may be dusted directly onto the grid or they can be suspended in a suitable liquid medium and then applied onto the support film.

SELECTED BIBLIOGRAPHY

Baumeister, W. and Hahn, M.: Specimen supports. In Hayat, M.A. (Ed.): *Principles and Techniques of Electron Microscopy: Biological Applications*, vol. 8. New York, Van Nostrand, 1978.

Bradley, D.E.: The preparation of support films. In Kay, D.H. (Ed.): *Techniques for Electron Microscopy*, 2nd ed. Philadelphia, F.A. Davis, 1965.

Bradley, D.E.: Techniques for mounting, dispersing, and disintegrating specimens. In Kay, D.H. (Ed.): *Techniques for Electron Microscopy*, 2nd ed. Philadelphia, F.A. Davis, 1965.

Hall, C.E.: *Introduction to Electron Microscopy*, 2nd ed. New York, McGraw-Hill, 1966.

Horne, R.W.: The examination of small particles. In Kay, D.H. (Ed.): *Techniques for Electron Microscopy*, 2nd ed. Philadelphia, F.A. Davis, 1965.

Meek, G.A.: *Practical Electron Microscopy for Biologists*, 2nd ed. London, Wiley, 1976.

Chapter 11

SHADOW CASTING AND REPLICA TECHNIQUES

THE TYPE OF SPECIMEN that can be examined in the transmission electron microscope (TEM) is limited by two factors—resolution and contrast. The resolving power of a TEM is usually determined by the instrumental design and other factors (*see* Chap. 5) and, in a way, is independent of the specimen. Nevertheless, to achieve this resolution in practice, one must have suitable specimens, and their preparation for examination in the TEM is the main problem of the microscopist. Differential scattering gives rise to contrast (*see* Chap. 4), and the specimen must be thin enough to transmit electrons. The resolution is dependent on the thickness of the specimen, and generally the resolution cannot be better than approximately one-tenth the specimen thickness. Thicker objects would scatter some electrons and produce contrast, but their resolution would be poor (*see* Chap. 3).

Granted that the specimen is thin or small enough, the greatest problem with biologic specimens is not resolution but *contrast*. Due to their low physical density (approximately 1), these specimens simply fail to scatter enough electrons out of the beam to produce a detectable image, i.e. they lack contrast. Therefore, the biologist usually cannot examine any living organism in the TEM because the dehydration and ionizing radiation would suffice to kill it. Second, all biologic specimens require some forms of preparation to adapt them to the instrument. Very small objects would have a poor contrast, and the resolution of thick or bulky specimens would be poor. The small objects can be detected (their contrast artificially

of the specimen illuminated obliquely by white light (Fig. 11-2b), the micrograph can be printed as a negative or an *intermediate reversal negative* can be made.

An ideal shadowing metal should have the following criteria: (a) it must be inert thermally and chemically; (b) it must scatter electrons readily; and (c) it must form a uniform and structureless deposit. The resolution, naturally, depends on the thickness of the deposit and the metal used. Metals with higher melting points that require high voltages for evaporation (*electron beam evaporation*) provide the best resolution. Metals with low melting points, on the other hand, are easily evaporated by the *resistance heating method* at low voltages in a conventional vacuum evaporator (*see* Fig. 10-4). However, they tend to form aggregates under the beam in the microscope and as such are not suitable for high-resolution shadowing.

An ideal shadowing requires the following conditions: (a) the source should be essentially a *point source*; (b) the evaporant should be pure; (c) the deposit should be of optimal thickness; (d) the dispersion of the evaporant should be symmetrical and spherical; (e) the lines of flight of the evaporant should be linear without elastic collision; and (f) the shadow contour should represent the exact geometrical projection of the specimen profile.

Shadowing Metals and Evaporation by Resistance Heating

A number of metals with higher density have been used for shadowing: gold, gold-palladium (60:40), platinum, platinum-palladium (80:20), platinum-iridium (80:20), platinum-carbon (60:40), chromium, zirconium, thorium, uranium, germanium, tungsten, iridium, tantalum, and molybdenum. Gold, which has been used for shadowing since 1946, is no longer used because it produces a fairly large granular background after electron bombardment. Gold-palladium, however, is used for coating scanning electron microscopic specimens. Chromium, due to its low physical density, requires a thick deposit for providing enough contrast and is also no longer used.

Platinum-carbon, platinum, and platinum-palladium are recommended for shadowing by the resistance heating technique. Of these, platinum-carbon gives the highest resolution. Low melting metals can be evaporated onto the specimen by a variety of procedures in a conventional vacuum evaporator. A *V*-shaped tungsten wire, a straight tungsten wire, a coiled tungsten wire, or a tungsten basket can be used as the filament. A molybdenum boat may also be used for evaporating the metal. Most metals, e.g. platinum, platinum-palladium, are available as wires (0.008″ dia.). The amount of metal to be evaporated depends on the resolution required and the size of the specimen. Normally, a small piece of metal wire (approximately 1 cm) can be wound around a straight tungsten wire to form a small loop. The metal is evaporated by passing a high current through the filament.

The vacuum pressure in the unit must be high (10^{-4} to 10^{-5} torr), and a number of evaporators are commercially available. The higher the vacuum, the finer is the metal deposit. A liquid nitrogen cooling system between the bell jar and diffusion pump should be used to minimize contamination. A vacuum loss usually occurs when the tungsten filament is first heated (degassing). It should, therefore, be heated to a temperature where the metal is seen to melt. The filament is then cooled down and reheated for final evaporation after the vacuum is reestablished.

The specimen grids can be mounted on a glass slide for shadowing (*see* Chap. 10), and they should be placed 10 to 15 cm away from the source. The angle of shadowing (α) varies with the specimen, and the larger the specimen, the larger is the angle. Small objects such as viruses usually require an angle of 10 to 15°. No apparent damage to the specimen occurs because the amount of hot metal deposit is very small. Electron micrographs of bacteria shadowed with platinum-palladium are shown in Figure 11-3. An apertured screen can be placed between the specimen and source (Fig. 11-4) to provide a sharper definition of the shadow and for protecting the specimen against the thermal radiation. In this method, the specimen can be placed as close as 4 cm to the source, and the technique apparently improves the shadowing resolution, especially with platinum-carbon.

The thickness of the metal deposit can be determined by placing a glazed porcelain with a drop of vacuum oil on it (*see* Chap. 10 for evaporation of carbon). Too heavy a metal deposit may damage the specimen, decrease the resolution, and increase the size of the

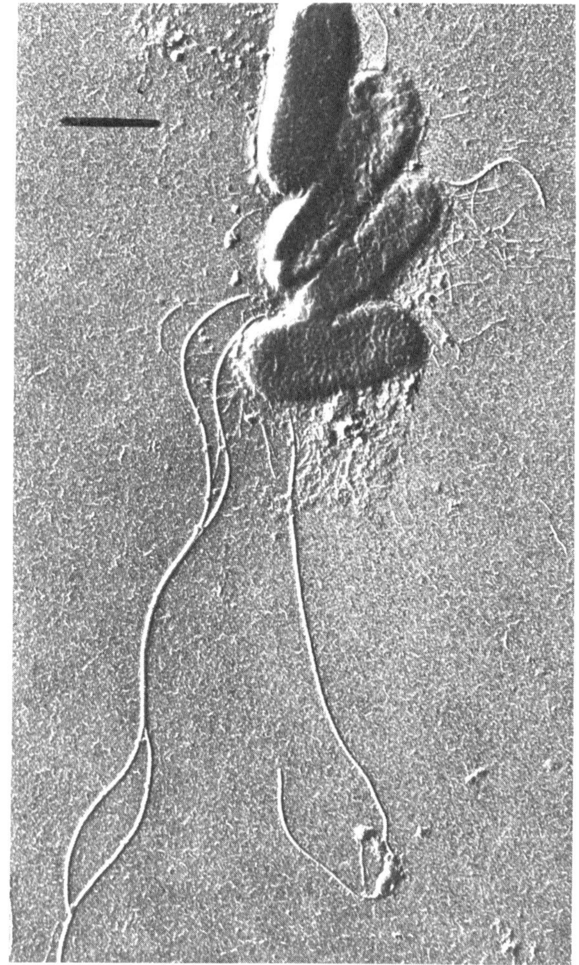

Figure 11-3. Bacteria shadowed with platinum-palladium. (a) *Proteus vulgaris*. Most of the flagella have been dislodged during preparation. Positive print. Bar = 1 μm.

specimen. Although the exact amount of metal to be evaporated is not critical, it can be roughly estimated by the following formula:

$$M = \frac{4\pi t r^2 d}{\sin \alpha}$$

where M = mass of metal (g); r = distance (cm) from the source to the

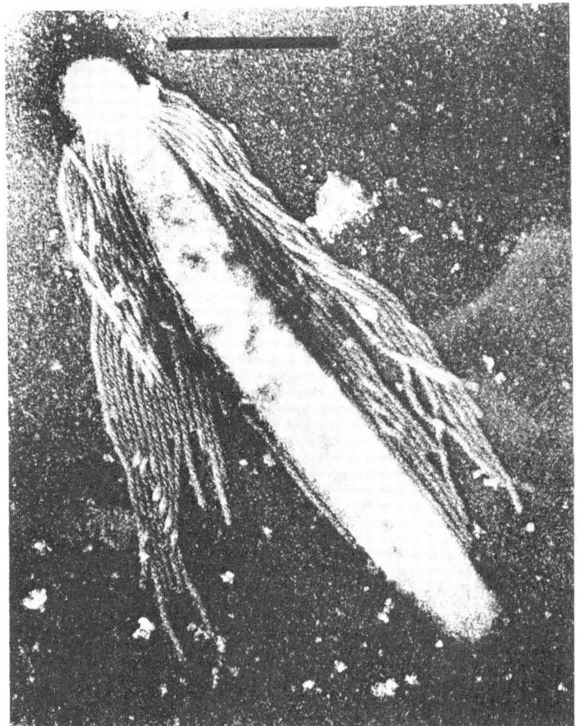

Figure 11-3. (b) Novel bacterium infecting an African snail. Characterized by a head with long, thick, flagella-like filamentous organelles. Negative print. Bar = 1 μm. From R.M. Cole, C.S. Richards, and T.J. Popkin, Novel bacterium infecting an African snail, *Journal of Bacteriology*, *132*:950, 1977.

specimen (perpendicular height of the filament); t = thickness of deposit (Å); d = density of the metal (g/cm^2); and α = shadowing angle.

The thickness of deposit can be calculated as follows:

$$t(\text{Å}) = \frac{M \sin \alpha}{4\pi r^2 d}$$

These formulas, however, are not accurate because the metal is not evaporated spherically as assumed, and the formulas may be off by a factor of two in some cases. The height of a particle can be calculated as follows:

$$H = \tan \alpha \times L$$

where α = shadowing angle; L = length of the shadow.

Other Shadowing Procedures

A variation of the shadow-casting technique is the *rotary shadowing* or *conical shadowing* introduced by Kleinschmidt et al. in 1962. It has been extremely valuable for transmission electron microscopy of nucleic acid molecules. Essentially, the metal is deposited onto the specimen at an angle while the specimen is being rotated and a motorized rotor is commonly used. The evaporant is uniformly deposited in all directions, and the contrast depends on the thickness and distribution of the evaporated film. Structures that might not be visible due to the their unfavorable orientation become visible due to the buildup of metal, which increases the diameter of the nucleic acid strands (Fig. 11-5).

In the *portrait shadowing* technique, the object is shadowed from two directions. After the initial shadowing, the specimen is turned through 180° and another film, approximately 20% of the original deposit is applied. Thus, the surface topography of what was initially the shaded area of the object can be examined, and a three-dimensional reconstruction can be made. The icosahedral structure of some viruses was determined by this method.

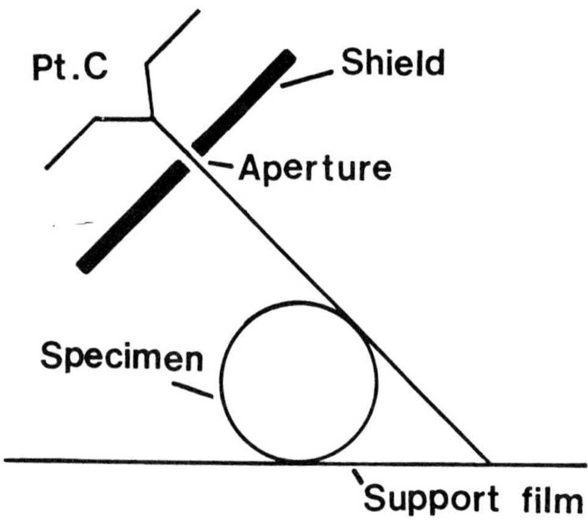

Figure 11-4. A shielded apertured screen placed between the specimen and source during metal evaporation.

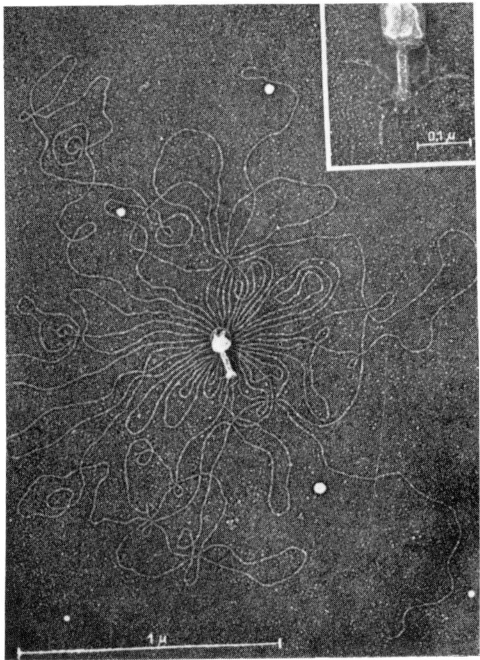

Figure 11-5. An osmotically shocked T_2 bacteriophage showing its DNA embedded in cytochrome c monolayer, rotary shadowed. Negative print. An intact phage is on the upper right. From A.K. Kleinschmidt, D. Läng, D. Jackerts, and K. Zahn, Darstellung und Langmessungen des gesamten Desoxyribonucleinsäure-inhaltes von T_2-Bakteriophagen, *Biochimica et Biophysica Acta. 61*:857, 1962.

Simultaneous evaporation of platinum-carbon provides the highest resolution (20Å) by the conventional vacuum evaporation technique. Although platinum-carbon rods or pellets are available commercially, a simple method for evaporating this mixture is to place a loop of platinum wire between the points of two carbon rods and apply a high current. Carbon (melting point (MP) ≈ 4000° K) does not melt but evaporates at the vapor pressure. The density of carbon is low (≈2), and a thick deposit is usually needed for adequate contrast. It is often difficult to obtain reproducible deposits of platinum-carbon for optimal results.

High-Resolution Shadowing

The resolution achieved by shadowing with low melting metals has been poor, and the technique has been replaced by the negative staining method (*see* Chap. 12) to a large extent. However, the shadow-casting procedure has now been improved considerably to provide a good resolution, and with the popularity of the freeze etching technique, shadow casting has created renewed interest in biology.

Problems in High-Resolution Shadowing

Although a deposit of 8 to 10Å thick heavy metal film provides sufficient contrast, particles in this size range cannot be resolved. This is due to the granularity of the shadowing film, which in actuality is a collection of isolated particles. When the atoms of an evaporated metal are deposited on the specimen surface, they release their excess energy until they maintain a thermal equilibrium with the surface. The atoms retain enough energy during this time to be highly mobile on the surface of the specimen. Before coming to rest, they would continue to wander a few angstroms unless they are immobilized by a very cold specimen surface, by strongly binding with the specimen, or by combining with other atoms or atom clusters of the evaporated metal.

These very small, stable clusters of atoms, called *nuclei*, would continue to grow during the evaporation. This is called *preferred nucleation* or *decoration effect* and can be misleading in a highly magnified image. The shadowing resolution, then, depends on the distance between these clusters, and the surface mobility of the evaporant has to be reduced to obtain a higher resolution.

The surface mobility can be suppressed and the background granularity reduced by keeping the specimen surface very cold. Unfortunately, when the temperature drops to −100° C or lower, the residual water vapor contaminates the specimen surface by ice condensation. A high-melting metal that has strong internal binding sites may also be used for shadowing to reduce the surface mobility of the evaporant. Alternatively, it can be achieved by simultaneously evaporating a high-density low-melting metal, e.g. platinum (MP 2,043° K), and a substance with a very high melting point, e.g. carbon.

The resolution can also be reduced by specimen damage (due to thermal radiation) and contamination during evaporation. Oil vapors arising due to back streaming from the vacuum system and grease or dirt cause contamination of the specimen. This can be substantially reduced by using cold traps or baffles, ion or turbomolecular pumps. The evaporation time should be as short as possible (≈10 seconds). When the bell jar is opened to the diffusion pump, the pressure should drop rapidly to 10^{-5} torr. Prolonged pumping causes oil contamination. The specimen can also be shielded against contamination.

Electron Beam Evaporation of High Melting Point Metals

The so-called *refractory metals* with MP above 2500° K have the finest grains, but they cannot be evaporated by resistance heating at a low voltage. These metals often require an evaporation time of 10 minutes to 10 hours in a conventional evaporator, consequently causing radiation damage and contamination. Tungsten forms an alloy with other low-melting metals. For high-resolution shadowing only platinum-carbon can be evaporated by resistance heating.

High melting point metals such as iridium (MP 2,727° K), molybdenum (MP 2,890° K), tantalum (MP 3,270° K), tungsten (MP 3,680° K) and tantalum-tungsten that provide high resolution (10A) are evaporated at a high voltage in an electron beam evaporator (Fig. 11-6). Basically, an electron beam is directed onto the material to be evaporated. The rods of high-melting metals are used as targets (anode), and the metal evaporates from the hanging drop that is formed at the tip of the anode. Ion deflection plates in front of the evaporator are used for protecting the specimen against bombardment with charged particles. In the past, a considerable amount of specimen damage was reported with commercially available electron beam evaporators. It now appears that with a proper design, electron beam evaporation can provide high-resolution shadowing. Although tungsten is susceptible to chemicals and cannot be used for shadowing freeze-fractured replicas, tantalum-tungsten alloy is highly stable to chemicals.

THE REPLICA TECHNIQUE

This technique permits an indirect observation of the surface structures of electron-opaque objects in the TEM. It was originally developed for examining surface structures of metals. Although the

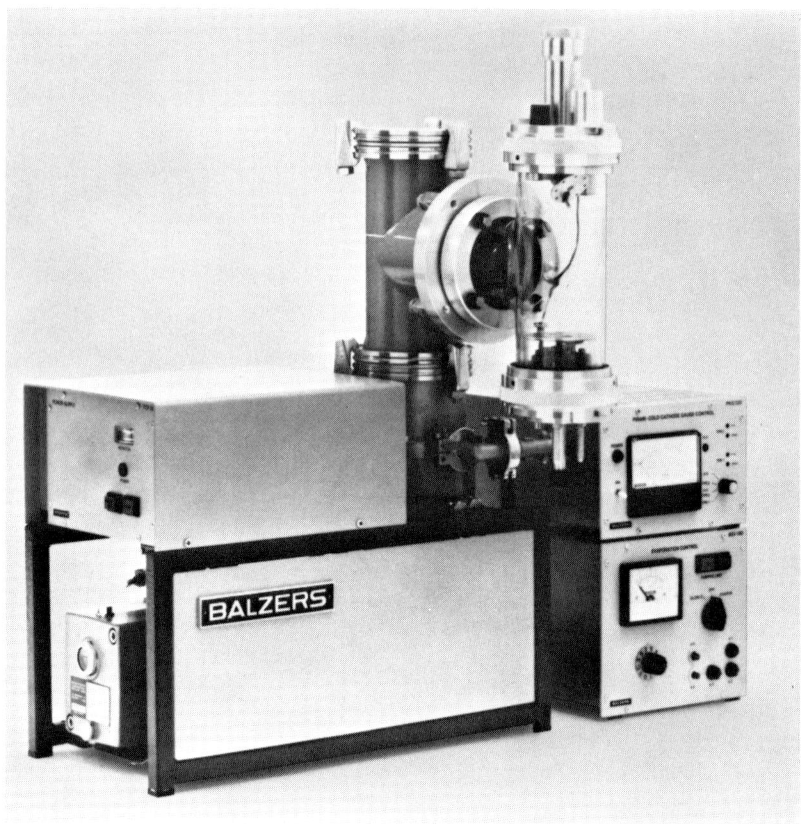

Figure 11-6. A compact vacuum evaporator with both conventional resistance heating and electron beam source evaporation facilities. Courtesy of Blazers Union.

technique is regularly used by metallurgists, it has also been employed for transmission electron microscopy of biologic specimens, although scanning electron microscopy has largely replaced this procedure. The freeze-etching method, a variation of the replica technique, has become very useful to biologists (*see* Chap. 15).

A replica consists of a thin film of an electron-transparent material that corresponds exactly to the surface topography of the specimen. Collodion was first used in 1940 for replicating surfaces of steel, and formvar was later employed as a replicating material.

Evaporated carbon film, introduced in 1954 by Bradley, was a major development in preparing replicas, and replication with carbon is now most widely used.

There are two basic methods. In the simplest *one-step* (*single-step*) replica technique (Fig. 11-7), the replicating material (plastic or evaporated carbon) is applied directly onto the specimen, and the replica is floated on water, collected on uncoated grids, shadowed to increase contrast, and examined. When a plastic replica is picked up by placing a filter paper on the grids, the surface topography of the specimen is reversed (*negative replica*). Plastic replicas from tissue surfaces are difficult to remove, and replicating with carbon is preferable for such specimens.

For the *two-stage* (*double stage*) replica technique (Fig. 11-8), an impression of the specimen is first made with a plastic, after which carbon is evaporated onto this impression. The plastic is dissolved and the replica is collected, shadowed, and examined. Often, a piece of cellulose acetate tape softened in acetone is applied onto the specimen surface. When it is hardened, it is first carbonized and then shadowed, or it is shadowed with platinum-carbon, which also acts as a replica. Finally, the cellulose sheet is dissolved in acetone, and the replicas are collected on grids.

The specimen can also be shadowed first and then replicated. This is used for viruses and other small particles, which can be sprayed on a clean glass slide or onto the surface of a freshly cleaved mica. It is then shadowed with a heavy metal and replicated by carbonization;

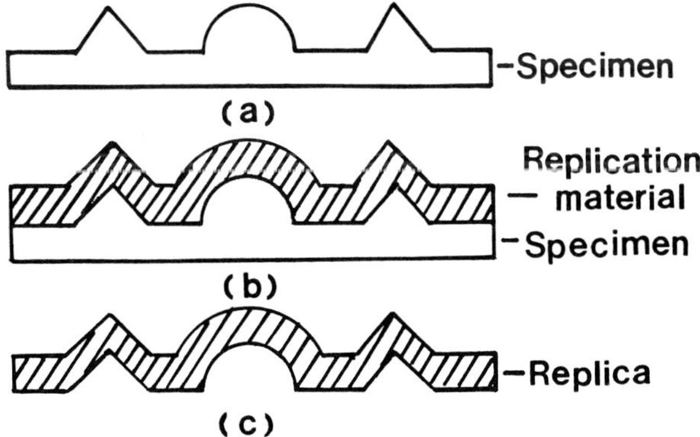

Figure 11-7. One-step replica technique.

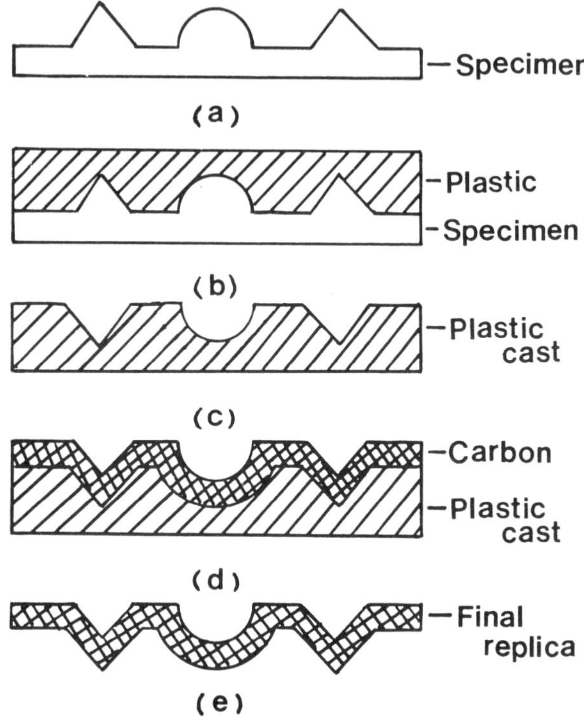

Figure 11-8. Two-step replica technique.

the replica is stripped onto the water and collected on grids. The resultant replica is called a pre-shadowed replica, and the technique is referred to as *shadow transfer technique* because, although they depict the surface topography as does any other replica, the picture appears like any shadow-cast specimen. A stereo pair of surface replica of an insect scale prepared by this method is shown in Figure 11-9.

Tissue Replicas

Replicas of tissues have been prepared by using water-soluble plastic, polyvinyl alcohol (PVA). It causes less damage to tissues, its thickness can be varied, and it can be easily removed without causing disintegration of tissues or damage to the replica. In this

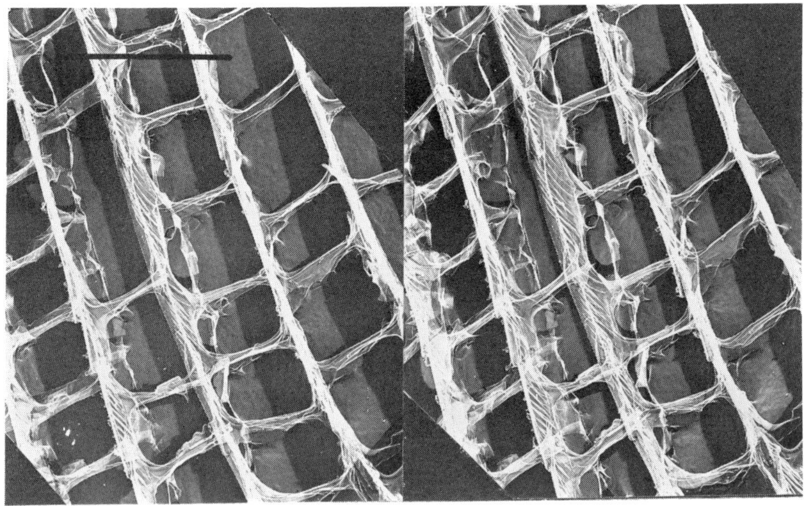

Figure 11-9. A stereo pair of surface replica of a small portion of a single scale of silver-spotted skipper butterfly. Bar = 1 μm. Courtesy of Dr. R. L. Steere.

procedure, a cellulose acetate sheet moistened in acetone is placed on a glass slide, and the tissue is embedded on it for easy handling. After hardening, the tissue may be trimmed to provide a better surface for replication. Strips of cellophane tape are applied around the tissue for making a shallow well, and 5 to 10% PVA is applied.

The PVA is allowed to solidify slowly overnight after which the film is coated with platinum-carbon. The replica is stripped off by dissolving PVA in a water bath at 97° C for 15 minutes. The replica is collected on grids and examined. Excellent results have been obtained with a variety of biologic specimens (Fig. 11-10), and this has now created renewed interest in replica techniques.

SELECTED BIBLIOGRAPHY

Abermann, R., Salpeter, M.M., and Bachmann, L.: High resolution shadowing. In Hayat, M.A. (Ed.): *Principles and Techniques of Electron Microscopy: Biological Applications*, vol. 2. New York, Van Nostrand, 1972.

Bradley, D.E.: Replica and shadowing techniques. In Kay, D.H. (Ed.): *Techniques for Electron Microscopy*, 2nd ed. Philadelphia, F.A. Davis, 1965.

Hall, C.E.: *Introduction to Electron Microscopy*, 2nd. ed. New York, McGraw-Hill, 1966.

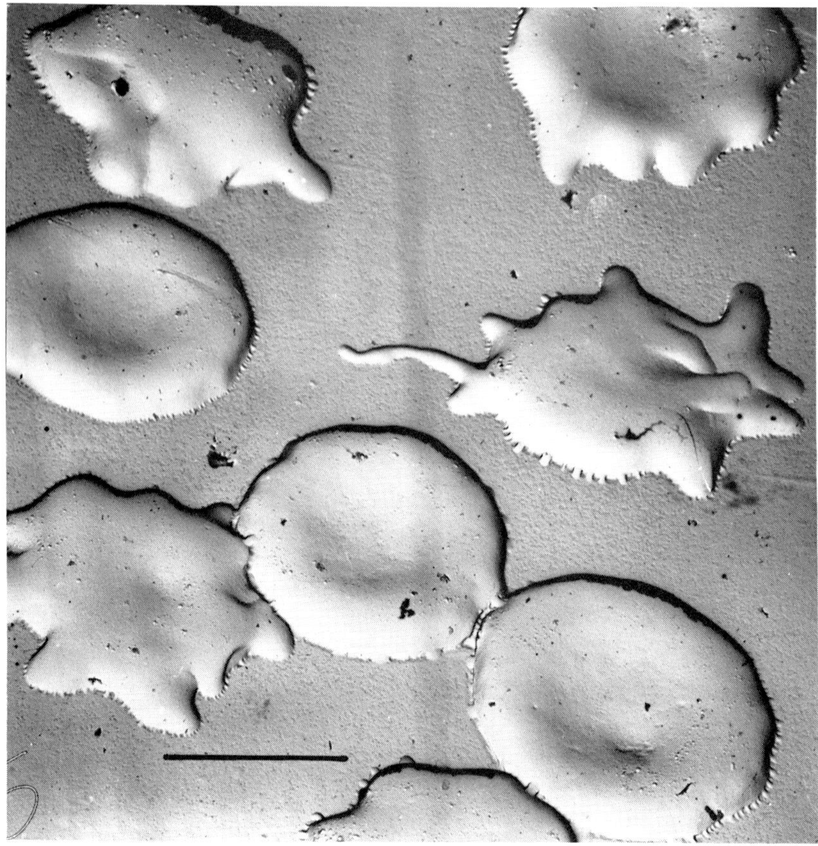

Figure 11-10. Tissue replicas. (a) Human acanthrocytes. Bar = 5 μm. Courtesy of Dr. W. J. Henderson.

Henderson, W.J. and Griffiths, K.: Shadow casting and replication. In Hayat, M.A. (Ed.): *Principles and Techniques of Electron Microscopy: Biological Applications*, vol. 2. New York, Van Nostrand, 1972.

Kleinschmidt, A.K., Läng, D., Jackerts, D., and Zahn, K.: Darstellung und Langmessungen des gesamten Desoxyribonucleinsäure-inhaltes von T_2-Bakteriophagen. *Biochim Biophys Acta*, *61*:857, 1962.

Preuss, L.E.: Shadow casting and contrast. In Bahr, G.F. and Zeitler, E.H. (Eds.): *Quantitative Electron Microscopy*. Baltimore, Williams and Wilkins, 1965.

Williams, R.C. and Wyckoff, R.W.G.: Application of metallic shadow casting to microscopy. *J Appl Phys*, *17*:23, 1946.

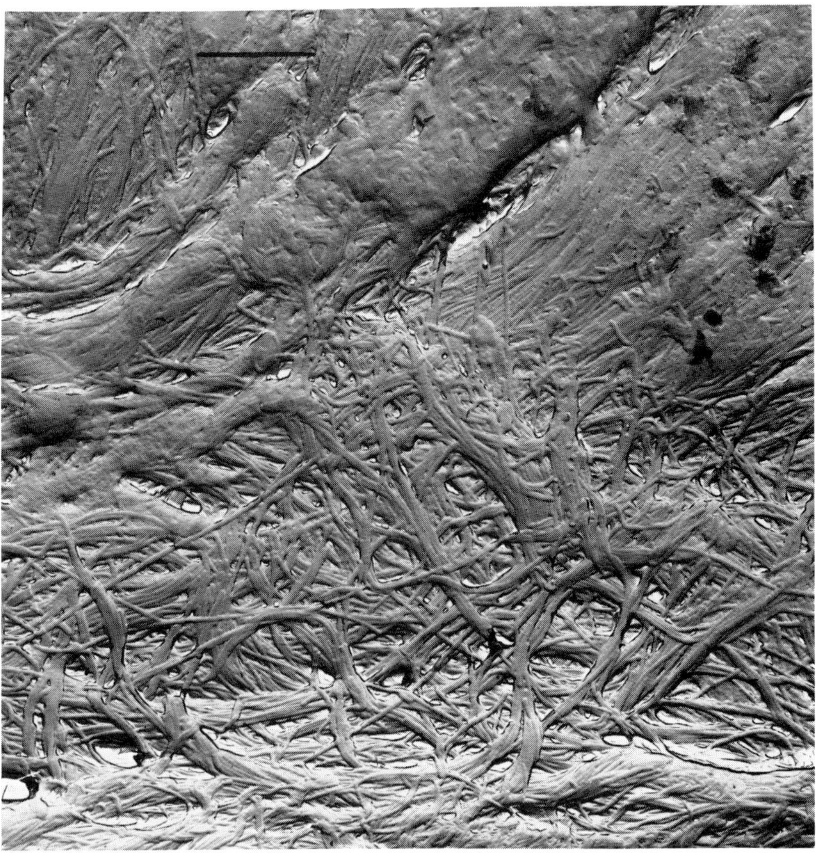

Figure 11-10. (b) Connective tissue of human mesothelioma. Bar = 2 μm. Courtesy of Dr. W. J. Henderson.

Chapter 12

NEGATIVE STAINING

IN THE PREVIOUS CHAPTER, we have discussed that it is not the *resolution* but the *contrast* that is more important in transmission electron microscopy of biologic specimens. It is generally recognized that for an object to be clearly visible in the transmission electron microscope (TEM), the product of its thickness (in Å) and its weight density (g/cm^3) should be more than 400. For an object to be visible at all, this product should be more than 100. Biologic specimens have a density of approximately 1, and for any biologic structure less than 100Å in size, some type of treatment for artificially increasing the contrast is necessary.

In addition to shadow casting, the positive staining technique may be applied to electron-transparent particles. This involves treatment with heavy metal stains of high atomic numbers, e.g. uranyl acetate, lead citrate, and phosphotungstic acid, and a staining time of a few minutes to several hours may be necessary. The stained particles appear as very electron dense objects against a relatively transparent background (Fig. 12-1a). The technique is routinely used for staining ultrathin sections (*see* Chap. 13), but it is unsuitable for studying structures at the subparticle level in the TEM.

The negative staining method involves surrounding or *embedding* the particles with a structureless thin layer of material of high weight density. When an object is embedded in such a stain with more than twice its own density, it will be seen with an increased but *negative* or *reverse* contrast. The objects thus appear as light areas or *holes* against a dark background (Fig. 12-1b) because of their lower electron scattering power than the stain. The technique reveals ultrastructural details of biologic macromolecules, e.g. viruses and

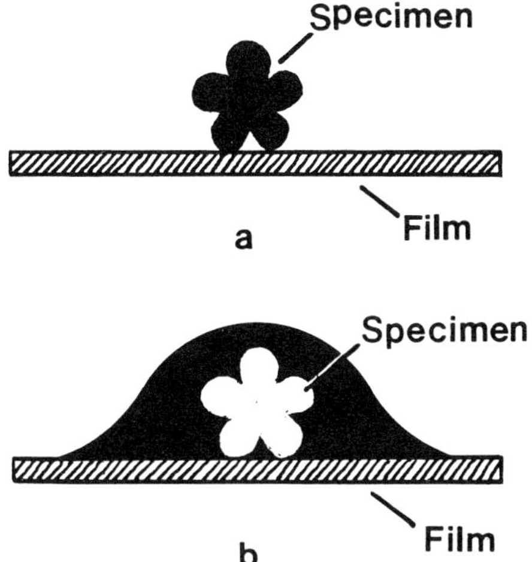

Figure 12-1. Diagrammatic representation of staining of small particles. (a) Positive staining. (b) Negative staining.

permits measurement of small molecular and intramolecular dimensions.

The negative staining technique has been used in light microscopy as early as 1892 but the effect of *reverse contrast* was first observed by Farrant in 1954 during electron microscopy of ferritin molecules. While observing bushy stunt virus particles positively stained with phosphotungstic acid (PTA) at pH 4.6, Hall, in 1955, reported that the preparation also had some negatively stained virus particles. In 1956, Huxley demonstrated that PTA penetrated the 40 Å diameter central hole in tobacco mosaic virus rods that was predicted to exist by x-ray diffraction analysis. The negative staining technique was standardized, and a simple spray method for routine use was developed by Brenner and Horne in 1959.

The negative staining technique is very popular among biologists because (a) it is very simple and quick; (b) there is good preservation, contrast, and resolution; and (c) only small volumes of specimen and stain are necessary. However, the technique is limited to the study of isolated particles, and a high particle concentration is necessary.

An object will not be negatively stained if the stain penetrates its structures and blacks it out. Apparently some protein molecules have open structures that are penetrated by PTA so that they are not visible in a negatively stained preparation. The loss of many substances into the background by penetration of the stain gives this technique an important advantage for examining virus particles. Very dirty preparations that are unsuitable for shadowing or positive staining, e.g. virus-infected tissue culture fluid, can be examined by the negative staining method. The debris penetrated by the stain is blacked out in the background, but the virus particles, into which the stain does not penetrate, stand out with reverse contrast.

PROPERTIES OF A GOOD NEGATIVE STAIN

The criteria of a good negative stain are as follows: (a) it must have a high weight density to provide a high contrast; (b) it must be sufficiently soluble in water (at least 80g/100 ml) so that it does not come out of solution immediately but only in the final stages of drying; (c) it should not react with the specimen and should have no staining affinity for it; (d) it should have a high melting point so that it is insensitive to the beam; and (e) its molecular size should be sufficiently small to penetrate into particle irregularities, it should spread in a thin layer (not thicker than the specimen), and the solidified stain should be nongranular.

Owing to the requirement that a good negative stain be sufficiently soluble and have a high anhydrous weight density, only a few stains have been found suitable as negative stains. These are PTA, sodium silicotungstate, uranyl acetate, uranyl oxalate, uranyl formate, ammonium molybdate, silver nitrate, and cadmium iodide. Among these stains, PTA has been most popular. It preserves structural details in most systems, and the presence of relatively high concentrations of salts such as phosphate buffer can be tolerated with this staining. Other popular stains include uranyl acetate, sodium silicotungstate, and ammonium molybdate.

Mechanism and Conditions of Negative Staining

The mechanism of this staining is not fully understood, but it is probable that the stain dries more rapidly than the specimen and solidifies into a smooth glassy film around the specimen. A final dehydration of the specimen occurs after the stain solidifies, and this is apparently responsible for preservation of the specimen to a large extent. Structural alteration is frequently observed when the specimen is dehydrated without embedding in the stain.

A correct mixture of the stain and particles is required for a good negative staining. Too much negative stain would give rise to large dense areas or droplets with very few visible particles, or the particles may be obscured completely. Conversely, insufficient stain would result in large electron-transparent areas around the electron-dense material. Particles with very little negative stain could collapse, and their true size cannot be measured. A negative staining can only be achieved if adequate wetting occurs. A wetting agent such as bovine serum albumin or preferably bacitracin may be added for adequate spreading of the particles on the support film (*see* Chap. 10). The concentration of the stain may also influence the appearance of the specimen.

A negative stain solution contains ions that may be absorbed onto the specimen, causing a change in fine structures and some positive staining. During drying, the combined charge effects of the support film and the droplet may cause dissociation or disruption of biologic specimens under certain conditions. The pH of the stain is very important, and it affects staining. The pH of a 2% PTA solution, for example, is highly acidic (<2), and it would stain positively unless the pH is adjusted to near neutrality. A volatile buffer such as 2% ammonium acetate or 2% ammonium carbonate may not be suitable for some specimens. Although a considerable change in the salt concentration and pH occurs as the specimen is dried down on the grid, the stain apparently causes no serious damage. Little morphologic distortions occur in negatively stained virus preparations, and most virus particles are often still infectious after mixing with PTA, or in some cases after mounting on the grid.

It is emphasized that a negative stain suitable for one specimen may not be suitable for another. The effect of the stain on the particle under study must be examined whenever possible. It is also important to examine the specimen immediately after staining, otherwise residual moisture would cause the stain to penetrate the specimen slowly. In some negatively stained preparations, only the top or bottom of the specimen is seen; both sides are visible in others.

Preparation of Negatively Stained Specimens

Negatively stained preparations require a high optical magnification and a high resolution for their examination. An anticontaminator and a double condenser system are mandatory for minimizing contamination and for recording fine details. The astigmatism must be compensated, and thin foil self-cleaning objective apertures should be used if possible. Although some microscopists prefer bare carbon films, formvar-carbon films (*see* Chap. 10) are satisfactory for this purpose. For best results, the specimen should be purified, concentrated in a density gradient, and dialyzed. An unbuffered staining solution is usually used. The pH of the most commonly used negative stain, PTA, must be adjusted to near neutrality with normal KOH or NaOH (called KPTA and NaPTA, respectively) to provide a good negative staining. The neutral stain is often simply called PTA without specifying if it is potassium or sodium salt, and no difference is usually detected between the two salts.

A 1 to 2% solution of PTA in a pH range of 6.5 to 8 is commonly used. Some microscopists have also used a 3 to 4% PTA solution. Although some workers prefer a freshly prepared solution of PTA, the stain is stable over a fairly long period of time, provided the pH does not drop. The pH of the unbuffered solution should be checked and adjusted if necessary before use. Uranyl acetate at a concentration of 0.2 to 1% is generally used. The pH of this freshly prepared stain is approximately 4.5, which is suitable for most works, and the required pH can be adjusted. Fifteen to thirty minutes are required for dissolving uranyl acetate in water. The solution is stable only for a few hours in the dark, and it should be freshly prepared before use. The usual pH range for ammonium molybdate is 7 to 7.4. The specimen can be embedded in a negative stain by a variety of procedures.

Spray Method

Brenner and Horne (1959) mixed equal parts of a virus suspension with 2% KPTA at pH 7.5 and atomized the mixture onto carbon-coated grids with a vapornephrin glass nebulizer. A high velocity spray gun may also be used for this purpose.

Drop Method

This is the simplest procedure routinely used in many laboratories.

The grid is held at the edge with a self-clamping forceps. A small drop of the specimen suspension is placed on the grid with a fine Pasteur pipette to form a *bead.* The droplet is touched with a filter paper to remove most of the excess liquid (some microscopists do not remove the excess fluid also). A drop of stain is immediately placed before the sample is dry. After allowing 30 seconds to 1 minute, the stain is blotted off by touching with a filter paper and the grid is ready to be examined. The sample may also be mixed with the stain (a ratio of 1:1 is commonly used) and can be applied onto the grid as a single drop.

Float Method

In this method, the sample is placed as a drop on the grid or the grid is floated upside down on a small volume of specimen suspension, e.g. a few drops on a depression slide or on a piece of parafilm. An appropriate time is allowed for the particles to adhere to the support film, and the grid is transferred to float on a few drops of the staining solution. The grid is then dried by touching with a filter paper and examined. Some particles adhering to the support film may be dislodged when the grid is floated on the stain.

Pseudoreplica

The specimen, in this technique, is placed on a solidified agar in a petri dish, evenly spread with a glass rod, and allowed to settle onto the agar for 15 to 30 minutes. The agar is then cut out, placed on a glass slide, and coated with 0.25 to 0.5% formvar solution in ethylene dichloride for replication. After the plastic film dries, the film is floated off onto a solution of uranyl acetate or PTA. Grids are placed on the top of the film and picked up by placing a filter paper on them.

Fixation

The specimen may be fixed for 15 to 30 minutes with 0.5 to 2.5% glutaraldehyde or 10% reagent grade formalin in phosphate buffer before negative staining. Superior preservation of tertiary and quaternary structures has been reported for some particles with this method.

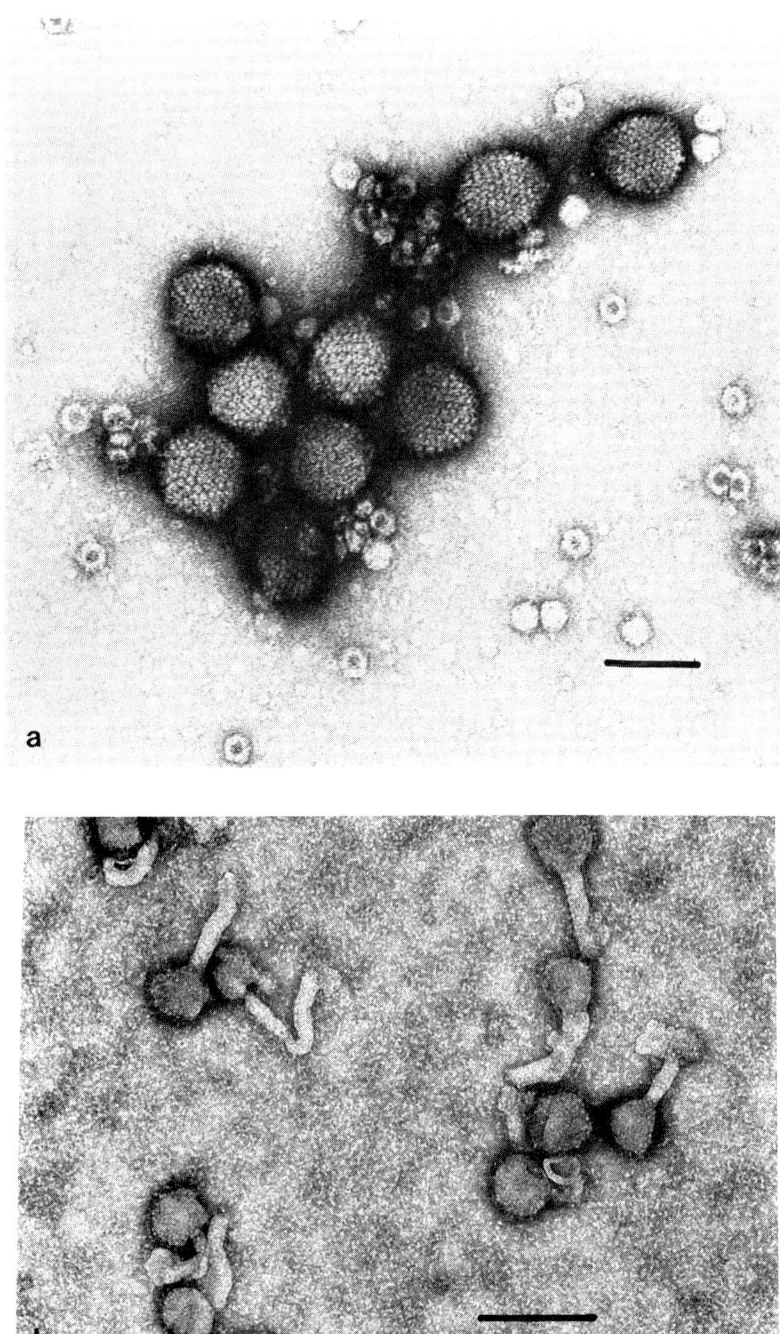

Figure 12-2. Virus particles negatively stained with 2% phosphotungstic acid. (a) A human adenovirus with its adenoassociated virus particles. Bar = 100 nm. Courtesy of Mr. G. Thomas. (b) Rauscher leukemia virus. Bar = 200 nm.

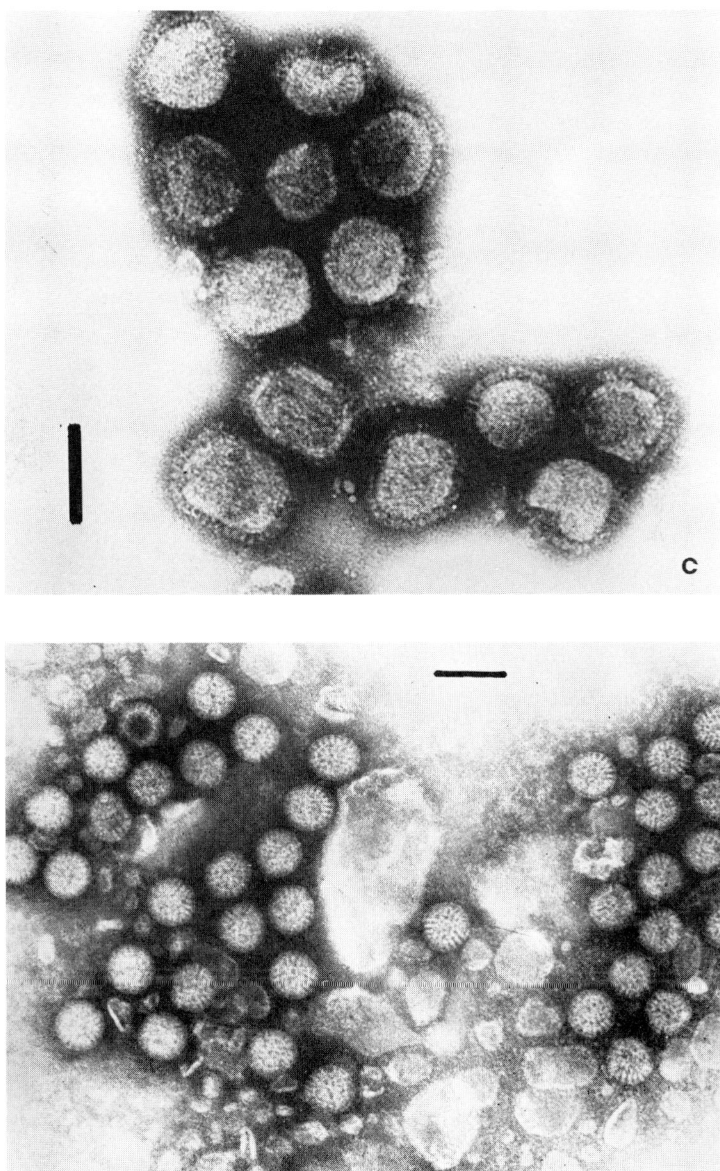

Figure 12-2. (c) Human influenza A virus. Bar = 100 nm. (d) Bovine rotavirus. Bar = 100 nm. Courtesy of Dr. S. McNulty.

Resolution and Interpretation of Negatively Stained Particles

Although a resolution of 7 to 10Å has been achieved under ideal conditions with some negatively stained preparations, a resolution close to 20Å is obtained with most specimens. The highest resolution can be obtained with uranyl stains because of their smaller size (4 to 5Å), which also causes an increased penetration of these stains. However, their contrast is somewhat lower (their density is 3.7) than tungstates, which give a somewhat lower resolution (their size is 8 to 9Å). Tungstates are usually less penetrating and can be used for specimens such as viruses when extensive penetration might obscure surface details and capsomere structure.

Measurements of particle size from electron micrographs should be approached cautiously because of the possibility of drying artifacts. Owing to the background structure especially present due to defocusing at a high magnification, particles in the size of 10Å or smaller should be interpreted with caution.

Application of the Negative Staining Technique

The technique has been applied to a wide variety of biologic specimens, but most extensive studies have been made with viruses (Fig. 12-2). Information obtained on viral ultrastructure has been extremely useful for classification of these agents. Structural features suggested by other physical and chemical techniques have been confirmed by ultrastructural study of some viruses, protein crystals, and bacterial flagella that illustrate the range and validity of negative staining.

Suspensions containing large organisms are difficult to stain negatively. A large amount of stain tends to penetrate into these specimens and their components. However, bacterial flagella and fimbriae are stained negatively (Fig. 12-3). These organisms should be washed in distilled water or a suitable volatile buffer, and care should be taken not to damage them during centrifugation. Subcellular components, collagen and muscle fibers, enzymes, phospholipids, frozen sections, and cell fragments from viral-infected cell cultures can also be examined by this technique.

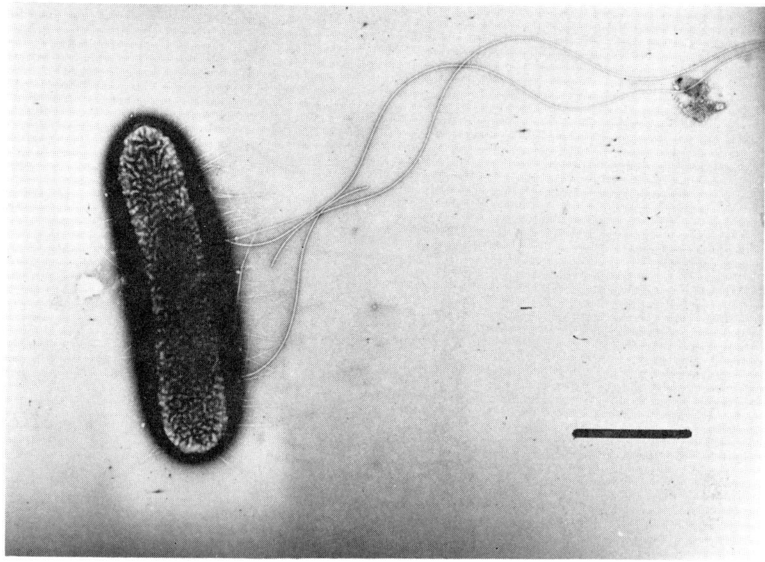

Figure 12-3. *Proteus vulgaris* negatively stained with 2% phosphotungstic acid. Note the negative staining of bacterial flagella and fimbriae. Most of the flagella have been dislodged. Bar = 1 μm.

Serologic reactions combined with negative staining can be applied for demonstrating specific antigen-antibody reactions on the virus surface (*see* Chap. 16). Virus particles in stool filtrates can be detected by negative staining, and along with immune electron microscopy (*see* Chap. 16), they are valuable in diagnosing neonatal diarrhea in man and animals. The technique is also useful for counting virus particles in a sample (*see* Chap. 14).

Optical aids for enhancing periodic details of some objects can be used with negatively stained electron micrographs. A suitable optical diffractometer can be used to extract the diffraction pattern of the photographic negative. A three-dimensional reconstruction of the image can be made by filtering out the nonperiodic background detail and noise (*see* Chap. 18). Superimposition methods can also be used for enhancing periodic details in negatively stained images.

SELECTED BIBLIOGRAPHY

Farrant, J.L.: An electron microscope study of ferritin. *Biochim Biophys Acta, 13*:569, 1954.

Glauert, A.M.: Factors influencing the appearance of biologic specimens in negatively stained preparations. In Bahr, G.F. and Zeitler, E.H. (Eds.): *Quantitative Electron Microscopy*. Baltimore, Williams and Wilkins, 1965.

Gregory, D.W. and Piria, B.J.S.: Wetting agents for biological electron microscopy. I. General considerations and negative staining. *J Microscopy, 99*: 251, 1973.

Hall, C.E.: Electron densitometry of stained virus particles. J Biophys Biochem Cytol, *1*:1, 1955.

Hall, C.E.: *Introduction to Electron Microscopy*, 2nd ed. New York, McGraw-Hill, 1966.

Haschemeyer, R.H. and Meyers, R.J.: Negative staining. In Hayat, M.A. (Ed.): *Principles and Techniques of Electron Microscopy: Biological Applications*, vol. 2. New York, Van Nostrand, 1972.

Horne, R.W.: The application of negative staining methods in quantitative electron microscopy. In Bahr, G.F. and Zeitler, E.H. (Eds.): *Quantitative Electron Microscopy*. Baltimore, Williams and Wilkins, 1965.

Horne, R.W.: Negative staining methods. In Kay, D.H. (Ed.): *Techniques for Electron Microscopy*, 2nd ed. Philadelphia, F.A. Davis, 1965.

Valentine, R.C. and Horne, R.W.: An assessment of negative staining techniques for revealing ultrastructure. In Harris, R.J.C. (Ed.): *The Interpretation of Ultrastructure*. New York, Academic Press, 1962.

Chapter 13

THIN SECTIONING AND ASSOCIATED TECHNIQUES

LONG BEFORE THE EXISTENCE of the transmission electron microscope (TEM), histologic techniques for light microscopy had been developed to a high degree of excellence. These techniques are similar to those used in electron microscopy, but modifications are necessary at each stage because the aim of electron microscopy is to observe fine structures. They must be preserved by suitable preparative procedures. Preparation of ultrathin sections for electron microscopy is still somewhat of an empirical art, and certain methods have given what are considered to be *good results*. A novice to this field must carefully study the aesthetically pleasing electron micrographs published by acknowledged and skilled electron microscopists and try to repeat them.

The principal aim of this chapter is to discuss the principles and the rationale of the TEM histologic techniques. A few most commonly used recipes for this purpose are given here. The reader is referred to books dealing exclusively or almost exclusively with thin sectioning for others.

The basic steps in preparing ultrathin sections are fixation, dehydration, embedding, ultramicrotomy, and section staining. In the early days of electron microscopy, specimen damage was frequently observed where light microscopic techniques were used, but this damage was not visible in the light microscope (LM). However, x-ray diffraction studies of periodic objects, freeze-etching, and negative-staining techniques have now proved beyond a reasonable doubt that the currently published cellular ultrastructures are true representatives of the cell in life, at least at the resolution level of 20 Å currently achievable with thin sections.

The criteria of a good preparation for electron microscopy are as follows: (a) The appearance of the specimen after fixation and embedding should be compatible with that seen in life. A 0.5 μm - thick section stained with toluidine blue should have no distortions that could be detected by phase contrast light microscopy of the living cell. If gross damage is visible with the LM, it is unlikely that the fine structures will be well preserved. (b) The size and orientation of fine structures or periodic structures should be compatible with measurements made by other physical and chemical techniques, e.g. x-ray diffraction. If a procedure clearly preserves these structures it may also be adequate for other specimens. (c) All membranes should be continuous, with no sharp breaks. (d) There should be no empty spaces that are completely devoid of any granular material, especially between the membranes. Although there is some controversy on this criterion, in *well-preserved* electron micrographs, a grayish granular material is found in these spaces. In a living state, all cellular spaces are usually filled with an aqueous solution of colloidal protein and salts, and these should show some kind of a precipitated structure. Empty spaces usually give rise to the suspicion that the extraction of material and shrinkage has occurred.

FIXATION

Some specimen structures are inevitably altered during preparative procedures, and the extent of these changes can be compared with results obtained by a number of different methods. It is emphasized that a particular fixative and an embedding medium suitable for one tissue may not be suitable for another. Chemical fixatives are commonly used for fixations in electron microscopy, and some molecular alteration probably occurs with them. Several physical techniques have been used for preparing tissues to avoid these effects. We will discuss the chemical fixatives first.

The aims of fixation are (a) to preserve the cellular ultrastructure as lifelike as possible; (b) to protect these structures against disruption during embedding and sectioning; and (c) to prepare them for subsequent staining and examination in the TEM. Fixatives are generally very selective in fixing the cellular components. An ideal fixative should kill the cells quickly and cause minimum swelling and shrinkage. The speed of killing is dependent on the rate of penetration, and fixatives with low molecular weight (MW) are most effective.

As the fixative penetrates into the cell, it transforms the viscous colloidal protein of the cytoplasm into a solid transparent gel. A fixative should not denature protein, and coagulative fixatives such as alcohols are useless for electron microscopy. Noncoagulant fixatives, on the other hand, transform proteins into a transparent gel. They are called *additive fixatives*, because they chemically become a part of the protein they fix. Among the fixatives presently used, double fixation with glutaraldehyde and osmium tetroxide (OsO_4) has given the most satisfactory result (*see* in this Chap.).

Obtaining the Specimen and Methods of Fixation

This is the first step in preparing thin sections for electron microscopy. Cytolytic and postmortem changes occur very rapidly, and they greatly influence the appearance of cellular ultrastructures. Although it was previously recommended that the tissues be fixed within minutes after death, it now appears that the autolysis does not take place as rapidly as was presumed, and the fine structures may still be preserved several hours after death. However, it is emphasized that the cells should come in contact with the fixative as quickly as possible after their normal or experimental environment is altered, e.g. after death, surgical biopsy, removal of cultured cells or organisms. The central zone of the fixed tissue may show some autolysis. A variety of procedures are used for fixation of the specimen.

Fixation by Immersion of Small Pieces of Tissues

The tissues can be excised from small animals under light anesthesia or the animal may be killed by decapitation. The excised tissue should be sliced on a nonabsorbent surface with a few drops of the fixative. A new scalpel blade or razor should be used to excise the tissue into pieces no larger than 0.5 to 1 mm^3 in size to insure rapid penetration of the fixative. A toothpick may be used to transfer the small pieces of tissue into a short wide-mouth vial containing the fixative, and the fixation is allowed to proceed to completion. The specimens should be kept in the original vial until they are ready for embedding. Periodic agitation is recommended to enhance exposure to the fixative.

In Situ Fixation

In this method, the organ of the anesthetized animal (or a small animal killed by decapitation) is flooded with the fixative for a short time. Organs are generally covered with a capsule, and it should be removed before applying the fixative. A thin slice of the tissue is then removed, cut into small pieces, and transferred to a small vial containing fixative for completing the fixation.

If deeper parts of the organ are required for study, the organ may be incised immediately after applying the fixative for facilitating *in situ* fixation. The fixative may also be injected directly into this cavity for organs containing a lumen.

Vascular Perfusion

The introduction of a fixative through the vascular system involves a rapid distribution of the fixative throughout the tissues and is considered ideal for nerve tissues. However, fixatives are potent vasoconstrictors, and when used alone, many arteries become closed shortly after coming in contact with them.

In this method, the abdominal cavity of the anesthetized animal is opened, and the descending aorta is cannulated. A vasodilator such as sodium nitrite is injected slowly through the cannula followed by a small volume of warm balanced salt solution, and then by a fixative. The posterior vena cava (and occasionally the jugular) is opened to provide adequate drainage during perfusion. Following perfusion, the tissue is excised, sliced, and fixed in a vial of fixative.

A large volume of fixative may also be injected into a suitable artery as an alternative to vascular perfusion, but it is somewhat less reliable. Excellent preservation has been obtained with vascular perfusion, but the technique requires a large volume of fixative.

Handling Other Specimens

Although human tissues obtained by the needle biopsy method

have been used for electron microscopy, surgical biopsy materials are more desirable. The TEM histology applied to both biopsy and necropsy materials has been extremely useful in the diagnoses of some human diseases (*see* Chap. 19).

Specimens such as bacteria, protozoa, spermatozoa, cultured cells, and cell fractions are centrifuged to form a pellet, and this is treated as though it were a piece of tissue. Some small specimens can be centrifuged to form a pellet immediately after fixation. The speed of centrifugation required to form a pellet depends on the specimen. The pellet is cut up into small pieces either after fixation or during dehydration. Some specimens that may be damaged by high-speed centrifugation may be pelleted by centrifuging at a low speed and then embedded in soft agar.

Factors Affecting the Quality of Fixation

A number of factors are critical for a good fixation. These are as follows.

pH and Types of Buffer

The pH of the fixative is critical because acidification, swelling, and disruption of fine structures occur with unbuffered solutions. Most tissues are fixed at a pH range of 7.2 to 7.5. A pH of 6.8 is generally used for plant tissues and a still lower pH ($\approx$6) may be needed for bacterial nuclear material. An alkaline pH of 8 is more effective for protozoa and embryonic tissues.

The veronal acetate buffer of Michaelis was favored by Palade and was routinely used for a long time. However, it has now been replaced by most commonly used *S*-collidine and Millonig's phosphate buffers. Veronal acetate buffer cannot be used with aldehyde fixatives because it forms a substance of nonbuffering compound with these fixatives. Other buffers that have been used are arsenate, sodium bicarbonate, chromate-dichromate, and cacodylate buffers. The buffer should be nontoxic and not react with OsO_4.

A 0.2 *M* *S*-collidine buffer (pH $\approx$7.4) is made by dissolving 2.67 ml of pure *S*-collidine in 50 ml of distilled water. Nine ml of *N*-HCl is added to this solution. The buffer is diluted to 100 ml with water.

The Millonig's phosphate buffer (pH $\approx$7.3) consists of 2.26% (w/v) $NaH_2PO_4 \cdot H_2O$ in distilled water (solution A); 2.52% (w/v)

NaOH (Solution B), and 5.4% (w/v) glucose (Solution C). The solution D is made by adding 41.5 ml of solution A with 8.5 ml of solution B. The final buffer contains 5 ml of solution C and 45 ml of solution D.

Tonicity

The tonicity (osmolarity) can affect the appearance of the fixed tissue. Isotonic fixative solutions generally fail to prevent swelling, and a slightly hypertonic solution is apparently effective for this purpose. Nonelectrolytes such as sucrose and glucose, which were previously added to the fixative solution, decrease the rate of penetration. They also increase the extraction of cellular material during fixation and dehydration. Electrolytes, especially the divalent ions of Ca and Mg, increase the penetration of OsO_4 considerably, but they also produce granularity. If used, their concentration should be very low.

Concentration and Temperature of Fixative

A longer time of fixation is generally required with a low concentration of fixative that causes extraction of cellular material, shrinkage or swelling of the tissue. A higher concentration of fixative destroys enzymatic activities and cellular ultrastructures.

Higher temperatures generally enhance fixation, and a cold temperature is required especially for some fixatives such as $KMnO_4$. Immersion of the tissue in a cold fixative has been advocated for a long time, and the excised tissue is generally brought in contact with a chilled fixative (0 to 4° C). Fixation can be completed either at 4° C or at room temperature. The rate of cellular extraction and autolysis is minimized at a low temperature. A temperature of 40° C is required for fixation of some resistant specimens such as bacterial spores.

Duration of Fixation

The optimal duration for fixation of most tissues is not known. A series of fixation periods may be tried and varied from fifteen minutes to four hours at 0 to 4° C. A very short period of fixation would not withstand further processing of the tissues, and proteinaceous substances usually leak out of the specimen during a prolonged fixation. A fixation period of 30 to 90 minutes is commonly used.

Types of Fixatives

A number of fixatives have been used for electron microscopy and they differ in their rate of penetration and fixation of cell components.

Osmium Tetroxide

This is mistakenly called osmic acid, but the reagent is a nonelectrolyte and forms no salts. Its fixative properties were recognized almost a century ago, and its reaction with unsaturated lipids is evident by rapid blackening of the tissue. It was used as a fixative in electron microscopy in the 1950s. Palade in 1952 was first to obtain reproducible results with this fixative and veronal acetate buffer. It acts not only as a fixative but also as an electron stain that provides contrast, a major advantage over many other fixatives. It penetrates tissues very slowly, hardens the tissue slightly, and causes swelling of the tissue. A fixation time of 15 minutes to 2 hours at 4° C with a 2% solution in distilled water or phosphate buffer is usually sufficient.

Osmium tetroxide fixes lipids readily. When it is reduced by unsaturated lipids, *osmium black* along with some osmium oxide as a by-product is formed, which provides contrast in the specimen. It also fixes phospholipids and lipoproteins. It does not interact with most proteins, but a partial reaction probably occurs with sulphur-containing amino acids. Although OsO_4 may not react with proteins, it may still preserve them and contribute to the preservation of fine structures. The reagent apparently does not fix nucleic acids. Coalescence of DNA fibers into coarse aggregates that may be considered an artifact commonly occurs in OsO_4-fixed tissues. It does not fix carbohydrates, although starch and glycogen may be affected to some degree.

Osmium tetroxide solution should be handled with extreme caution. The chemical is volatile and poisonous, and its vapor fixes and kills the superficial epithelial cells almost instantaneously. It is especially harmful to the epithelial cells of the eyes, nose, and mouth. The reagent, both solid and in solution, should be handled with rubber gloves in a fume hood. When used on the bench, a fan must be arranged to blow air across the operator's hands so that the vapor does not come in contact with the face.

The chemical is expensive and is normally supplied as pale yellow crystals in sealed ampoules. A 2% solution, either in double distilled water or phosphate buffer, is commonly used for fixation. It is a strong oxidizing agent and is readily reduced in the presence of organic matter and light. The ampoule should be washed thoroughly to remove the label and organic glue. A clean uncontaminated glass-stoppered bottle should be filled with the correct amount of water or buffer to prepare a 2% solution. The ampoule is placed in the bottle, shattered with a heavy glass rod, and stoppered immediately. The crystals dissolve slowly, and a couple of days may be required for dissolution. Dissolution can be hastened by carefully heating the Pyrex® bottle at 60° C. The bottle should be wrapped with aluminum foil for protection against light and then refrigerated. The concentration of the solution decreases rapidly, and a small volume should be made at one time.

Aldehydes

Glutaraldehyde, formaldehyde, and acrolein (acrylic aldehyde) are commonly used fixatives in electron microscopy. Of these, glutaraldehyde is most popular as a primary fixative. Aldehydes are especially useful in cytochemical studies because, unlike OsO_4, they do not destroy enzymatic activities. Owing to their small MW, they generally penetrate the tissue rapidly, prevent gross distortion, autolysis, and bacterial degradation. They form gels with proteins by cross-linking and stabilize fine structures. Tissues fixed with them have to be postfixed with OsO_4.

Glutaraldehyde penetrates slower than other aldehydes and usually causes shrinkage of both the cell and nucleus. A fixation time of 30 minutes to 2 hours in a 2 to 10% solution in phosphate buffer is usually used. Overnight fixation, however, is not harmful, and this flexibility is one of the major assets of this fixative. The temperature of fixation is not critical, and a room temperature is generally used. Glutaraldehyde apparently increases the permeability of the tissue to the embedding media. A partial reaction of the fixative occurs with glycogen, and it also reacts with polyalcohols. It may react with nucleic acids, but it does not fix lipids.

Glutaraldehyde generally contains impurities such as glutaric acid, acrolein, ethanol, and various other polymers. A 25% solution of purified glutaraldehyde is usually supplied in ampoules that are stable for a long period of time at 4° C.

Formaldehyde (called *formalin* when diluted with water) is an indispensable fixative for light microscopy. Paraformaldehyde was previously used in electron microscopy. A 10% formalin in phosphate buffer prepared from a reagent-grade formaldehyde has been used as a fixative for TEM histology.

Acrolein is probably the most reactive among the aldehydes. It is an extremely hazardous chemical because it is flammable, highly toxic through the vapor and oral route of exposure, and is a strong skin irritant. A 5% solution in cacodylate or phosphate buffer along with a fixation time of 2 to 3 hours has been used for fixation of tissues.

Permanganates

Luft (1956) introduced buffered $KMnO_4$ to electron microscopy. It is a useful fixative for membranes, but it is a strong oxidizing agent. It dissolves cytoplasmic protein and introduces numerous artifacts. A 0.6 to 3% solution of $KMnO_4$ in veronal acetate buffer with a fixation time of 15 minutes to 2 hours at 4° C has been used. A freshly prepared solution should be used, and postfixation with OsO_4 is required.

Other fixatives that have been used in electron microscopy include ruthenium tetroxide, Dalton's chrome-osmium tetroxide, and $HgCl_2$.

Double Fixation

Among the fixatives currently used in electron microscopy, a double fixation with glutaraldehyde (*primary fixation*) followed by OsO_4 (*secondary or postfixation*) appears to be most effective. Glutaraldehyde fixes proteins but not lipids, whereas OsO_4 readily fixes lipids. Both fixatives may have a partial reaction with carbohydrates and nucleic acids. Therefore, when used together, they virtually fix all cell components. Glutaraldehyde, the primary fixative, penetrates the tissue rapidly and prevents autolysis, bacterial degradation, and gross distortion. The postfixative, OsO_4, penetrates slowly, fixes lipids, and provides added contrast. The tissue is generally fixed with 2.5% glutaraldehyde in phosphate buffer for 30 minutes to 2 hours, washed three times in the buffer, and then postfixed with 2% OsO_4 for 30 minutes to 1 hour.

Tissue Storage

Although biologic specimens are fixed, dehydrated, and embedded immediately, this may not be possible under unavoidable circumstances. The fixed tissue may be stored for a long period of time by the following method.

The tissue is fixed in 100% buffered formalin (1 part of reagent grade formaldehyde plus 9 parts of 0.05 *M* Sorenson's phosphate buffer, pH 7.2 to 7.4, for 1 hour at 4° C. The tissues are then placed at the room temperature for weeks or months (the longest time tested was 1 year) without changing the solution. Before postfixing with OsO_4, the tissue is washed in phosphate buffer overnight, dehydrated, and embedded in an epoxy resin. Although occasional artifacts are observed, the fine structures are almost identical to that observed in tissues directly fixed with OsO_4.

The Sorensen's phosphate buffer (0.2*M*) contains 2.78% (w/v) NaH_2PO_4 in distilled water (solution A), and 5.365% (w/v) $Na_2HPO_4 \cdot 7H_2O$ (solution B). The working buffer (pH $\approx$7.2) is made by adding 28 ml of solution A to 72 ml of solution B and then diluting to a total volume of 200 ml with water.

WASHING AND DEHYDRATION

The fixed tissue must be washed before dehydration to minimize a possible reaction between the fixative and the dehydrating agent. This is particularly important for double fixation, and the excess glutaraldehyde must be removed before postfixation with OsO_4. Otherwise, the two fixatives would react with each other, and a fine, dense precipitate of reduced osmium will be formed, and the specimen would be ruined. The washing should be done in the same buffer that was used for preparing the fixative. Generally two to three rinses for a total of 10 to 15 minutes are adequate. If 0 to 4° C was used for fixation, the washing should also be carried out at this temperature. Short wide-mouth vials with plastic caps are most satisfactory for tissue processing, and the fixation, washing, dehydration, and infiltration should be done in the same vial. The solutions can be withdrawn with a Pasteur pipette or poured out, but all solution changes must be done rapidly to prevent drying of the specimen.

Since embedding media are not miscible with water, they must be removed and replaced with a suitable organic solvent that is miscible in the embedding monomer. *Dehydration* is carried out by placing the tissue in increasing concentrations of alcohol or acetone. These solvents cause shrinkage and extraction of cellular components, especially lipids. Therefore, dehydration should be as short as possible to minimize these effects, but it must be complete. The tissue should be dehydrated in the cold if this temperature was used for fixation and washing. The specimen can be brought to room temperature during the last change in the solvent.

Traditionally, a graded series of ethanol (30%, 50%, 70%, 90%, and 100%) for about 15 to 30 minutes at each step has been used for dehydration. Although alcohol is miscible with methacrylate (which is no longer used for embedding), it is not readily miscible with resins. Therefore, an intermediate or *transitional solvent* such as propylene oxide, which is miscible in alcohol as well as in resins, must be used. After alcohol dehydration, the tissue is placed in propylene oxide for thirty minutes. This chemical is very reactive, extracts lipids readily, and is toxic and flammable.

Acetone apparently causes less shrinkage than alcohols, is miscible with all embedding media, and is very effective when rapid dehydration is needed. The tissue is passed in 30%, 50%, 70%, 90%, and 100% acetone (5 to 10 minutes each), and two changes of 100% acetone are used.

Inert dehydration with glycols has also been used (*see* in this Chap.). Automatic tissue processors for fixation and dehydration of a number of tissues at one time (Fig. 13-1) are now available. A few of these processors can also be used for infiltrating the tissues, e.g. Polaron and Sakura tissue processors.

Added contrast in the specimen can be obtained by staining the

Figure 13-1. An automatic tissue processor for fixation and dehydration of a number of tissues at one time. Courtesy of Microlab products.

tissue blocks during dehydration. This is called *block staining*. Uranyl acetate and PTA are not soluble in absolute acetone, but they are soluble in dilute solutions of acetone. Stains can be added to the final dehydrating alcohol bath. If PTA is used in the final alcohol bath, the PTA-containing alcohol should not be mixed with propylene oxide. Otherwise, an explosion and fire could result because of the reaction between alcohol and propylene oxide that is catalyzed by PTA. Block staining can also be done during washing after fixation.

EMBEDDING

Embedding enables the object to be cut sufficiently thin for examination in a TEM. The embedding medium must be hard enough to support and hold the tissue together.

Embedding Media

Paraffin and other harder waxes used in the early days of thin sectioning were too soft and unsuitable for the TEM. A good embedding medium should permit thin sectioning with the least damage during specimen preparation and the least interference during microscopy. The attributes of an ideal embedding medium are as follows: (a) it should be soluble in alcohol or acetone; (b) it should have a low viscosity for proper infiltration; (c) it should bond tightly to the tissue, harden uniformly without causing polymerization damage, and the polymer should be transparent; (d) it should be stable under electron bombardment; (e) its density should be as low as possible so that it would cause minimal electron scattering. Thus, the contrast of the specimen would not be reduced by the embedment; (f) the resin should not chemically alter the tissue; and (g) the polymerized resin should cut well. A number of embedding media have been used in transmission electron microscopy.

Methacrylates

The introduction of methacrylates by Newman *et al* (1949) was a major step toward preparing routine thin sections for the TEM. A mixture of n-butyl methacrylate and methyl-methacrylate (80:20) was commonly used. The monomer is miscible with alcohol, has very low viscosity and the polymerized plastic has good cutting qualities. Unfortunately, the plastic causes polymerization damage due to uneven polymerization so that stresses are set up that disrupt the fine structures of the specimen. The plastic shrinks up to 20 percent during hardening, is beam sensitive, and evaporates when irradiated. This, in a way, is good for contrast because the electron-scattering embedment is removed, but the cellular fine structures are collapsed and the evaporated plastic gives rise to contamination of the column. Therefore, methacrylates have now been replaced by epoxy resins.

Epoxy Resins

First introduced by Maaløe and Birch-Anderson (1956), the cross-linked epoxy resins are the most popular embedding media in transmission electron microscopy. They are characterized by the presence of epoxy end-groups, which are highly strained three-membered rings with the following general formula:

used as an embedment. The Vestopal-W embedding medium is prepared as follows:

Vestopal-W	—	100 ml
Initiator (benzoyl peroxide)	—	1 ml
Activator (cobalt naphthenate)	—	0.5 ml

It is important to mix the resin and initiator first and then the activator. An explosion will occur if the initiator and activator are mixed together first.

Water-soluble Embedding Media

A considerable interest for water-soluble embedding media has been created among electron microscopists ever since gelatin was used for this purpose in 1957. Gelatin blocks are very hard and difficult to cut. Water-soluble embedding media are useful for cytochemistry, particularly for enzymatic reactions. Since the tissue has to be dehydrated in alcohol or acetone, which is considered harsh for the tissue, it was presumed that water-soluble media would cause very little or no extraction of cellular materials. The tissue is usually dehydrated in increasing concentrations of these water-soluble media at 4° C and finally embedded in them. However, it is now recognized that these media are themselves organic solvents and cause extraction of cellular materials, particularly lipids. Furthermore, they are hydrophilic, the hardened blocks tend to absorb moisture, and the sections tend to swell on the water trough. Water-soluble embedding media include glycol methacrylates (2-hydroxyethyl methacrylate also called GMA, and hydroxypropyl methacrylate), aquon (extracted from Epon 812), Durcupan, and PolyAmph-10. They are commercially available (*see* Appendix for source of supplies). Aquon is prepared as follows:

Aquon	—	10 ml
DDSA	—	25 ml

Ten ml of this mixture is mixed with 0.1 ml of BDMA just before use.

Handling the Resins

Epoxy and polyester resins should be handled cautiously. Most of these monomers have been used simply because they have given satisfactory results, and their harmful properties have not been fully investigated. All these resins and the chemicals used to prepare the final embedding mixtures may cause contact dermatitis, other allergic reactions, or may even be carcinogenic. They are usually harmless after polymerization (hardening) in the oven. It is recommended that plastic gloves be used for their handling and the work carried out in fume hoods. All spilled resins should be wiped off with acetone. Disposable plastic wares and glass vials for tissue processing are recommended, and a covered wastebasket should be kept adjacent to the hood. It is emphasized that the resins and their chemicals require thorough mixing for insuring proper polymerization. The measurements can be done either volumetrically or by weighing. The monomer should never be poured in the sink. It is advisable to place the disposable plastic and glass wares containing leftover resin mixture in a 60° C oven for polymerization before discarding them.

Infiltration

This involves a gradual replacement of the dehydrating agent with an embedding medium. It is a critical process, and poor polymerization occurs if the tissue is not infiltrated with the resin. The tissue in absolute acetone or the transitional solvent propylene oxide is usually placed in 1:1 mixture of the solvent and the final resin mixture for 30 minutes to 1 hour at room temperature. This is followed by two changes in 100% plastic mixture (30 minutes to 1 hour each). A shorter infiltration time causes less extraction of cellular material, and the tissue is not subjected to osmotic damage. A mechanical agitator, e.g. a rotator or shaker, can be used for a short, uniform infiltration.

Embedding Procedure

Embedding consists of complete impregnation of the interstices of the tissue with the medium. The final embedding is best done in predried gelatin (size 00) or polyethylene capsules, e.g. Beem capsules (Fig. 13-2). Polyethylene Beem capsules with the ends already pointed into truncated four-sided pyramids are especially suitable because the block trimming is simplified with their use and

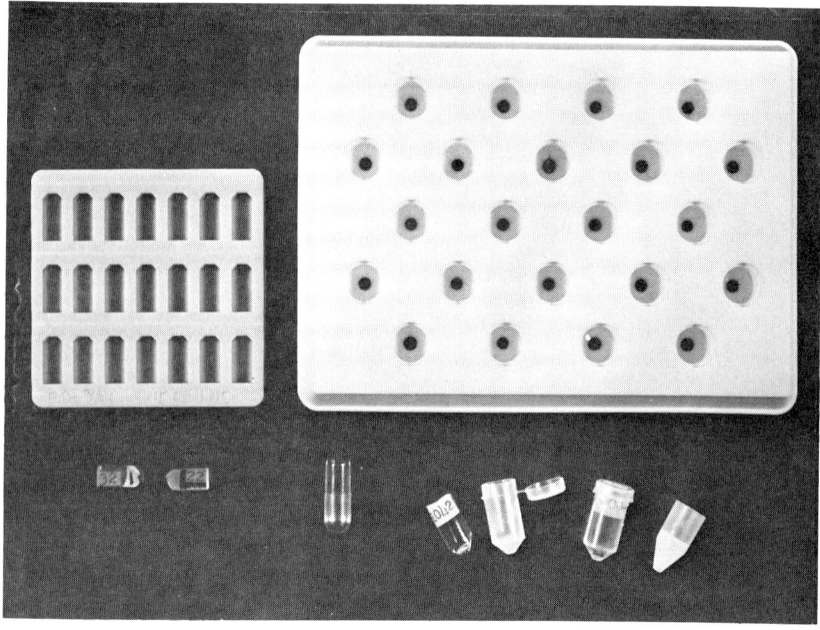

Figure 13-2. Gelatin and Beem capsules, Beem capsule holders, flat embedding silicone rubber molds, and embedded capsules.

the blocks are easily stripped after polymerization. Beem capsules of various sizes and shapes are available.

The capsules are placed in a capsule holder, and a small drop of the final embedding medium is placed at the base of the capsule. The tissue is transferred to the capsule with a toothpick, centered, and properly oriented. Permanent labelling of the embedded tissue is very important (Fig. 13-2). Records should be kept in a hardbound notebook, and serial numbers can be assigned to the specimens. A small piece of written paper can be rolled around the inside of the capsule at the top. The capsules are then filled to the top with additional resin and closed. The resin can be applied with a syringe, or it can be carefully poured from a plastic beaker. Care should be taken to eliminate air bubbles. Polymerization is accomplished after twenty-four to forty-eight hours in a 60° C oven. Tissue flotation before polymerization may occur, and the tissue should be gently pushed down to the base of the capsule.

Flat Embedding

Orientation of some tissues such as nerve or muscle fibers and teeth may be a problem because of their size and shape. A flat embedding is recommended for these tissues, and flat embedding silicone rubber moulds are available (Fig. 13-2). The flat embedding can also be done in a petri dish. A small portion of the polymerized resin containing the tissue is cut out and glued to a lucite rod for cutting thin sections. The flat embedding process is considered faster than capsule embedding.

Embedding Thin Layers of Cells

Although cell cultures are usually dislodged from the growing surface, centrifuged, and the resulting pellet then processed like a piece of tissue, cells grown on coverslips can be processed by the following simple method. The coverslip culture is fixed, dehydrated, and infiltrated while it is still in the Leighton tube. Before embedding, it is removed, and a capsule half filled with partially polymerized resin is inverted over a selected area of cells on the coverslip. A number of capsules can be placed in this manner. After polymerization, the coverslip is touched to a block of dry ice (solid CO_2) that easily detaches the coverslip, leaving the cell layer embedded in the block. The block should be placed in a desiccator to prevent condensation of moisture on the cool surface and then sectioned in the usual manner.

Rapid Embedding

This is particularly useful for surgical biopsy tissues for rapid diagnoses, and the entire process of tissue processing can be completed in three to four hours. Small pieces of tissue (0.5 mm^3) are fixed at room temperature in 2.5 to 5% glutaraldehyde and then in 2% OsO_4 (30 minutes each). The tissues are washed for 3 minutes after each fixation. They are dehydrated in acetone (30%, 70%, 95%, 100%, 4 minutes each). Two changes are given in the absolute acetone; the tissue is then placed in the acetone-resin mixture (1:1) for 15 minutes and finally in the resin mixture for two changes, 10 minutes each. Embedding with Epon (or its substitute) or with a low-viscosity epoxy resin is recommended. A mild agitation in a shaker or rotator also is recommended during the entire procedure. The tissues are then embedded (a flat embedding is preferred by some microscopists) and polymerized at 100° C for 1 to 2 hours. The blocks are ready to cut after cooling at room temperature. The process apparently causes no distortion of fine structures, and the results are comparable to those obtained by the conventional methods.

ULTRAMICROTOMY

The cutting of good, useful, and consistent ribbons of ultrathin section is the most difficult task in electron microscopy. The art of thin sectioning requires practice, patience, and dexterity in the highest degree. Successful sectioning depends on proper tissue preparation, a good microtome, a sharp, correctly adjusted knife, and last but not least, the skill of the operator. The basic principles of ultramicrotomy will be discussed here.

Block Trimming

All specimens, whether embedded in gelatin capsules, Beem capsules, or in a flat embedding mould, require some trimming prior to sectioning in the ultramicrotome. The end of the block containing the tissue should point into a four-sided truncated pyramid. Blocks embedded in polyethylene Beem capsules are ideal because they have this shape, require little trimming of the square face only, and they are easily stripped by making one longitudinal slit with a razor blade. Blocks embedded in flat embedding moulds are also easy to trim.

The gelatin coating can be chipped off; soaking of these capsules is not very effective. The cutting quality of the block apparently improves over time after initial curing, and it is a good practice to wait at least several days before trimming and sectioning the blocks. Individual blocks in a batch may vary in their cutting quality; this can be judged during trimming. Too hard a block usually shatters and the tissue would tear out; too soft blocks would give a characteristic rubbery feel during trimming.

The block is mounted in a chuck, and a dissecting microscope is used for trimming. The rounded end of the gelatin block is first cut horizontally across the top with a new razor blade (Fig. 13-3) deep into the specimen until a layer of plastic with osmicated black tissue comes off. It is advisable to examine this slice in a light microscope to insure that the proper tissue is being examined and for proper tissue orientation. Four successive cuts at 45° angles are then made. The truncated pyramid is then cut into a trapezoid by cutting off alternate sides of the block face until the largest side (the basal edge of the block) is approximately 0.2 mm. The two major sides of the trapezoid should be parallel and cut clearly to insure obtaining a straight ribbon during thin sectioning. The block face should be as small as possible and a good size is 0.1 to 0.2 mm^2. A large block face causes chatter and rapid dulling of the knife.

Block trimming machines are also available that include pyrimitome (LKB), Block shaper (Cambridge Instruments), and TM-60 (Reichert). The Beem capsule embedded blocks can be conveniently trimmed with the LKB ultrotome (a common practice in the author's laboratory).

Ultramicrotomes

An important contribution leading to ultrathin sectioning was made by Pease and Baker (1948) when they substantially reduced

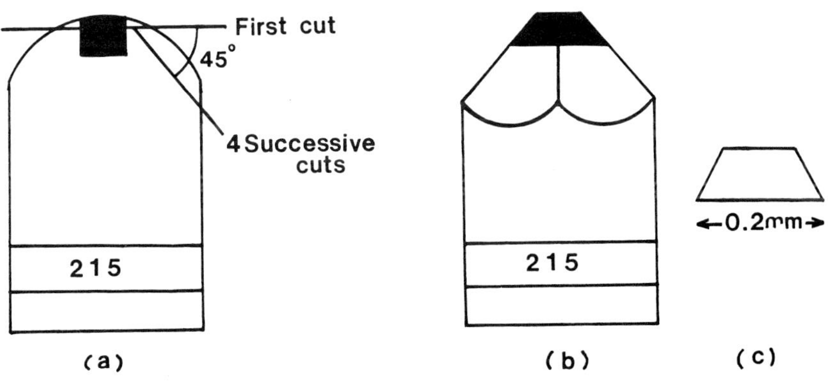

Figure 13-3. Trimming of gelatin capsule embedded blocks. (a) Side view of the stages of trimming. (b) Side view of the trimmed block. (c) The trapezoidal block face; the base of trapezoid should be ≈ 0.2 mm.

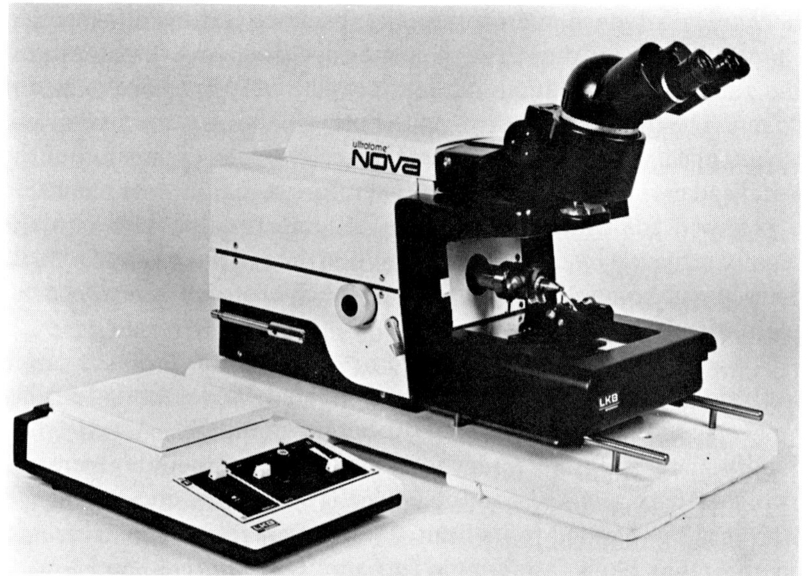

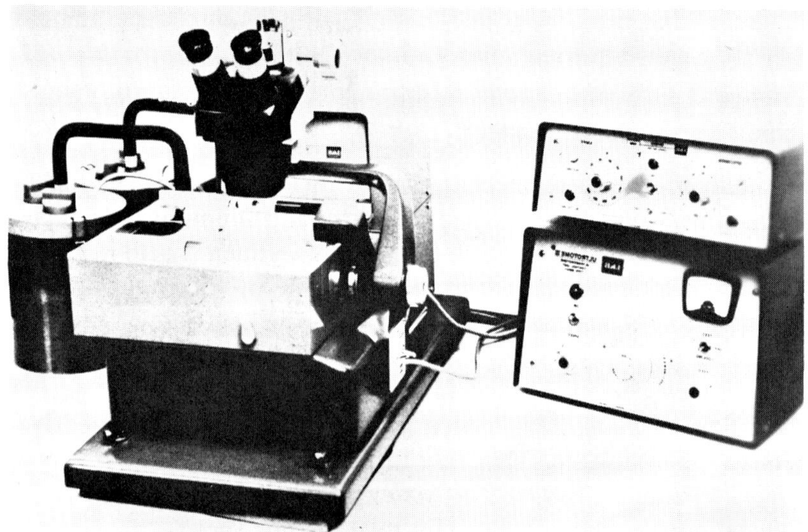

Figure 13-5. Commercial ultramicrotomes. (a) The LKB ultrotome Nova. (b) LKB cryoultramicrotome. Both courtesy of LKB Instruments, Inc.

has the most experience) is a versatile machine in which blocks embedded in Beem capsules can also be trimmed and thick sections cut by manual operation. The knife advance is manual, and the automatic advance is thermal. The arm carrying a fully orientable chuck moves only up and down. There is no bypass stroke, and the knife holder is retracted from the block face by excitation of an electromagnet.

Other thermal advance micotomes include the *Reichert Om U3* (Austria), an excellent fully automatic instrument.

Cryoultramicrotomes

Attempts have been made to cut thin sections of rapidly frozen tissues for electron microscopy similar to those cut on a cryostat for light microscopy. Attachments for *cryoultramicrotomy* are available for cutting frozen thin sections in conventional ultramicrotomes (Fig. 13-5b). The tissue is rapidly frozen in liquid nitrogen, and sections are cut in a conventional ultramicrotome in which the knife and the block are kept in a specially insulated chamber cooled by liquid nitrogen. Sections can be collected on grids from the dry knife or from a flotation medium such as cyclohexane. They are then kept in the cold microtome chamber for one hour or freeze-dried and examined either stained or unstained. Ice crystal formation causing fine structure disruption still appears to be a problem with this technique. The unstained sections are commonly used for x-ray microanalysis, and the sections are stained for enzyme cytochemistry.

Knives

An ultrathin section is apparently formed by splitting off of a surface layer of the specimen embedded in the resin. The cutting edge of the knife acts as a wedge, the effect being similar to that of an axe. The sections and the block suffer stresses during cutting. Metal knives are unsuitable for ultramicrotomy.

Glass Knives

First introduced by Latta and Hartmann (1950), glass knives are most commonly used for thin sectioning. They are sharp, easy to make, and inexpensive so that they can be discarded after a single use. They are made from plate glass and glass strips (16″ × 1″ × 0.25″), which are commercially available. Some workers prefer white hard glass; others like glass with a high silica content.

PREPARATION OF GLASS KNIVES. Glass knives can be prepared by breaking the glass with glazier's pliers or with special pliers commercially available for this purpose. More reproducible results are obtained with knife-making machines. Glass knives are made by the free-breaking technique regardless of the instrument used.

For making knives manually using special pliers, 1-inch wide glass strips are scored with a wheel type of glass cutter. The strip is scored to prepare 1-inch squares and broken by applying the exact pressure with the pliers. The square is then scored diagonally again, and two glass knives are made by breaking the square that has an angle of 45°. Knives can also be made from a rhombus, and the angle of these knives varies from 50 to 55°. Obtaining a good knife by this method requires considerable practice, and the results may be irreproducible at times.

Knife-making machines, e.g. LKB knife maker (Fig. 13-6), almost invariably produce two good knives every time, and a *knife maker* is an excellent investment for an electron microscope laboratory.

The finished glass knife (Fig. 13-7) has an arc or *fracture ridge* found below the edge of the back surface of the knife, which is formed due to stress generated during knife making. It serves as a guide to the best portion of the cutting edge, which is usually near the top of the ridge. A supercooled liquid, glass is subject to flow at room temperature and becomes dull. It is advisable to prepare glass knives just prior to sectioning. The number of good sections that can be cut with a glass knife may vary from fifteen to thirty.

EVALUATION OF GLASS KNIVES. The knife edge can be evaluated by darkfield illumination at a magnification of about 20× (Fig. 13-8a). The edge should be straight and even. The LKB ultrotome has a convenient system for inspecting the edge. An average knife has three zones (Fig. 13-8b). A really good part of the knife (Zone I, on the left) is scarcely seen and is the sharpest. The brighter the line, the poorer is its cutting power (Fig. 13-8a). Zones II (fair) and III (poor) on the right usually have *check marks* (*whiskers, frills*).

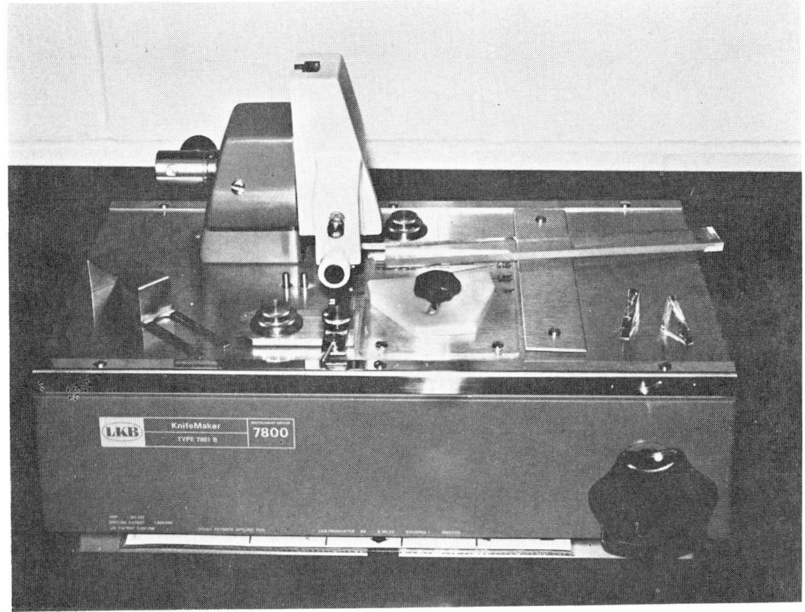

Figure 13-6. The LKB knife maker.

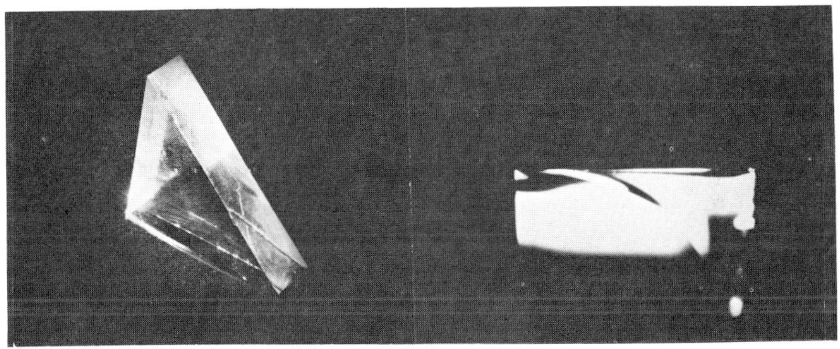

Figure 13-7. The glass knife (on the left) and the knife edge showing fracture ridge (on the right).

Diamond Knives

Diamond knives (Fig. 13-9) were introduced by Fernández-Morán in 1956, and they are commercially available, e.g. DuPont, LKB. They are extremely sharp, ideal for cutting hard materials,

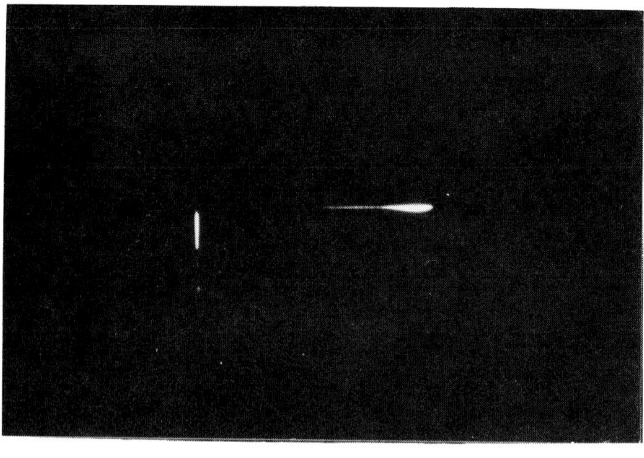

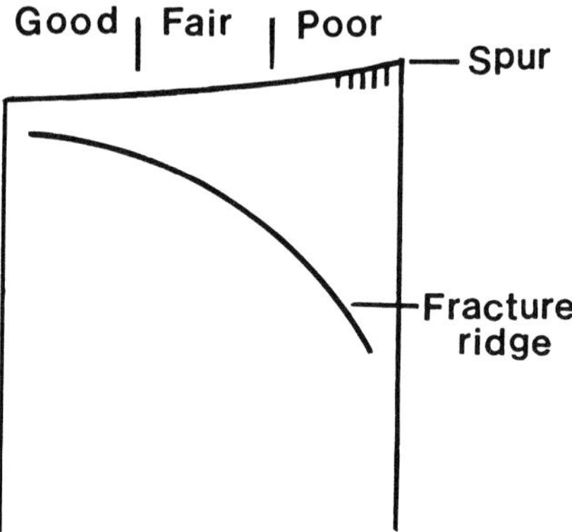

Figure 13-8. Darkfield examination of glass knives and the appearance of glass knives. (a) Darkfield evaluation of knife edge. (b) A diagrammatic representation of the cutting edge.

Figure 13-9. Knives with troughs. A glass knife with trough made of electrician's tape (on the left), a glass knife with a metal trough (in the middle), and a diamond knife with its built-in trough (on the right).

and have a long life span. In the author's experience, a diamond knife cuts more consistent ribbons of good sections than a glass knife, and some knives seem to last forever. Diamond knives are quite expensive, as much as $1,000/mm cutting edge. They exhibit extreme variation in their cutting quality even though obtained from the same source. Resharpening of these knives, done by the manufacturer, is also very expensive, and a resharpened knife may not cut satisfactorily.

The edge of the diamond knife is extremely brittle and can be easily damaged. A novice to the field should not cut with this expensive tool until he acquires sufficient experience in cutting with a glass knife. Diamond is extremely hydrophobic, and the knife edge is difficult to wet with water in the trough. The edge can be made hydrophilic by rubbing it with a detergent or saliva. A flattened wood toothpick may be used for this purpose, working along the entire cutting edge and allowing the cutting edge to split the wood.

The edge should be cleaned after use with a toothpick as described, and the toothpick can be soaked in acetone to help remove the material sticking to the knife edge. The knife is then washed in distilled water, dried with warm air, and stored in its container in a dry place.

The Knife Trough

Ribbons of ultrathin sections are extremely delicate and difficult to pick up on grids. Therefore, they are collected by floating them on water contained in a *trough* or *boat* attached onto the knife behind the knife edge (Fig. 13-9). Troughs can be made from masking tape or metals. Troughs prepared from masking tape are convenient because they can be discarded along with the knife.

Electrician and metallic tapes are also quite suitable for this purpose. A strip of tape is wrapped around the knife, and the excess is trimmed off. The tape should not overlap the cutting edge or the front face of the knife, and care must be taken not to touch the edge of the knife. The lower part of the tape is then sealed to the knife with melted dental wax. Metal troughs are good because they do not contaminate the trough fluid, but they require thorough cleaning after their use. Disposable plastic troughs are also available. Holders of diamond knives are made so that the trough is an integral part of the knife.

SECTIONING

Before cutting thin sections, the operator should be thoroughly familiar with the parts of the microtome and its controls. The microtome must be placed on a sturdy table, and an extreme stability of the microtome is necessary for vibration-free cutting. The table should not be placed in the path of air currents. Holders for different block sizes come with the microtome, and the block should be firmly mounted in the chuck. The fluorescent light is switched on, and the block face is properly oriented for cutting. The knife with its trough is placed in the knife holder and everything is made tight.

Distilled water is the usual trough fluid, but some workers prefer 10% acetone in this flotation fluid. A glass syringe should be used to apply water into the trough and the surface cleaned with a lens paper if necessary. The correct level of water in the trough touching the knife edge is very important, and the light should be adjusted to obtain correct reflection on water for judging the interference color of sections. If the water level is too low, the section would be compressed and stick to the edge of the knife. If it is too high, water would wet the block face and no sections would be cut.

The position of the knife relative to the block face is critical for cutting good thin sections, and three angles are involved in this setup (Fig. 13-10). The most important angle here is the *clearance or tilt angle*, α, between the cutting edge of the knife and the block face. It

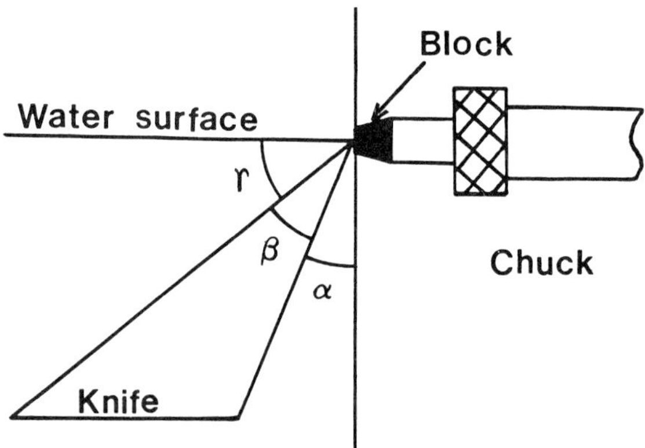

Figure 13-10. The knife, block, and cutting angles. α = clearance angle (tilt angle); β = knife angle (bevel angle); and γ = rake angle (sheer angle).

permits the specimen to clear the knife on the return stroke. The cutting edge is inclined in such a way that it is not parallel to the block face. Otherwise, the block face would scrape the back of the knife after cutting one section. A smaller clearance angle (usually 2 to 5°) is desirable, provided the block does not scrape the back of the knife after cutting. Instead of cutting thin sections, too large a clearance angle tends to scrape them, which results in *chatter*, and the knife may also be chipped.

The *knife angle* (*bevel angle*), β, varies from 45 to 55°. Knives with 50 to 55° angles are considered stronger. The knife angle plus the clearance angle determine the actual *cutting angle*. The *rake angle* (*sheer angle*), γ, is the angle between the line perpendicular to the front face of the block and the upper facet of the knife edge.

After everything is set up, the knife is advanced toward the block by using the fine knife advance control. The block face and the knife are very carefully watched, and it should be possible to see the reflection of the knife edge in the block face. The block face is then wiped off with a lens paper, and a few thick sections ($\approx 0.5\ \mu m$) are cut. Thin sections can now be cut manually, or the microtome can be adjusted to cut sections automatically (with an automatic machine). A cutting speed of 1 to 2 mm/second is desirable. Once some sections start to come off, the adjustments are reset so that pale gold colored sections are produced. The controls are then gradually adjusted so that gray and silver sections would dominate the ribbon.

The ribbon should be straight (Fig. 13-11), and the section should have no *knife marks*. These marks usually run perpendicular to the cutting edge. *Chatter*, which consists of alternate zones of thick and thin sectioning, is the most common bothersome section defect. They may not be detected during sectioning but are visible during electron microscopic examination. The knife edge can be cleaned with a flattened toothpick. During cutting, the sections are best viewed at a magnification of 7×.

Section Thickness

The thickness of sections can be determined by the interference colors produced by the sections as they float on water. The colors can be seen through the binocular with a properly adjusted reflected light. It should be pointed out, however, that the judgement of

Figure 13-11. Ribbon of sections floating on the water in the trough with correct illuminating conditions. Courtesy of DuPont Co.

interference color may not be sufficiently accurate because the eye does not notice thickness difference within the section, and the overall judgement varies from one person to another. The thickness of the sections is below the resolution limit of the eye, and their interference color is not visible to the naked eye. Sections of different thickness generally produce the following approximate colors according to the Peachy scale:

Gray	$<$ 600 Å
Silver	600 - 900 Å
Gold	900 - 1,500 Å
Purple	$>$1,500 Å

Thin sections are essential for the best resolution. Purple sections are useless, and gold sections usually give a poor resolution. The contrast, of course, decreases with thin sections, but after staining, they provide good contrast. Gray and silver sections are ideal for transmission electron microscopy.

Flattening the Thin Sections

When a ribbon of sections is cut, some sections may be slightly wrinkled, and their interference color cannot be judged. They can be flattened and their color determined by exposing them to the vapor of a solvent such as acetone or chloroform. A small brush dipped in the solvent is held a few mm above the floating sections. Care should be taken so that the brush does not touch the sections or the flotation fluid.

Mounting the Thin Sections

The sections are easily picked up by touching a clean grid (held at the edge with a forceps) to the sections as they float on water (Fig. 13-12). Some microscopists prefer to collect the sections on the matt surface of the grid. The sections would stick to the grid because they are hydrophobic. Bare (naked, uncoated) grids are commonly used, but grids with a formvar-carbon net can be used for this purpose. It is emphasized that epoxy-embedded sections collected on bare grids are very susceptible to the electron beam when first irradiated. A low-intensity beam should be used for a few seconds to stabilize them (curing). They can be then examined with a normal viewing intensity. Attempts should be made to place the sections as close to the center of the grids as possible.

Figure 13-12. The process of picking up sections floating on the water in the trough.

Alternatively, the sections can be picked up by bringing the grid up under the floating sections (*fishing technique*). The grids should be placed on a filter paper in a covered petri dish with section side up and allowed to dry.

The sections can be teased away from the knife edge and the ribbon broken into a few pieces with the tapered end of an eyelash taped to a toothpick or to an applicator stick. Serial sectioning is similar to the normal sectioning technique. Serial sections are required for a three-dimensional reconstruction of the image. Each and every section cut is needed and must be of the same thickness. The entire ribbon must be collected on the grid, and their mounting requires some skill. A single slotted grid (*see* Chap. 10) coated with formvar-carbon is generally used. The grid is positioned so that the ribbon lies directly under the slot. It is then touched down on the ribbon and the entire ribbon is picked up.

Problems Encountered in Thin Sectioning

Many factors may be involved in giving rise to faulty sections, and it is difficult to list all of them. The important factors, however, are faulty embedding, faulty cutting, and a lack of skill on the part of the operator. Ultramicrotomes are generally extremely reliable, and the machine usually plays a negligible role in producing faulty sections. Some common difficulties and their possible causes are listed below.

a. *No sections at all*—the advance mechanism has reached the end of its travel, knife may be too far back so that it does not touch the block face, ambient temperature change or air movement, incorrect clearance angle, knife edge dry due to insufficient water in the trough, or block and knife faces are wet.
b. *Skipping (sections cut on every other stroke)*—loose block, too large a block face, or too large a clearance angle.
c. *Sections vary in thickness*—dull knife, too small a clearance angle, soft block, loose block, or ambient temperature change.
d. *Knife marks*—knife edge dull, dirty, or chipped.
e. *Compressed sections*—dull knife, soft block, or loose block.
f. *Ribbon is not straight or no ribbon is formed*—top and bottom of the trapezoidal block face are not parallel, or too much liquid in the trough.

g. *Chatter (thick and thin bands across each section)*—This may not be detected at the time of sectioning. Knife and/or block loose, too large a clearance angle, dull knife, too fast cutting speed, block is too hard or ambient temperature change.
h. *Section pulled back by passing block*—knife or block face wet, soft block, or dirty block face.
i. *Folding of sections at the knife edge*—soft block, or dry knife edge.
j. *Dirty sections*—dirty trough and water, or dirty knife and block face.

Very few sections are completely flawless, and different parts of the section may have to be searched for a good photographic recording. Chatter marks, periodic variation in the contrast of section due to alternate zones of compression and zones of good sectioning, are especially annoying. They are often seen only during electron microscopy (Fig. 13-13), and they generally run parallel to the cutting edge while the section is being cut. Lipid bodies always show chatter due to local vibration at the time of cutting, which is caused by uneven localized polymerization. Knife marks (Fig. 13-14) usually run perpendicular to the cutting edge.

Section Staining

A 0.5 μm-thick section can be stained with an alkaline solution of 1% toluidine blue and examined in the light microscope. This is good for judging the quality of the preparation, and the exact area can then be selected for final block trimming.

The ultrathin sections for electron microscopy are positively stained with heavy metal stains for increasing the differential electron scattering power of different cell components. Epoxy resin-embedded sections are notoriously poor in their contrast. Epoxies, due to their high density, scatter electrons, which increases background noise, thus reducing the specimen contrast. The precise action of electron stains is not clear, and they seem to be somewhat nonspecific. The contrast of the specimen can also be enhanced by block staining during dehydration (*see* in this Chap.), and OsO_4 provides contrast to the specimen. A number of stains have been used for staining ultrathin sections, but a double staining with

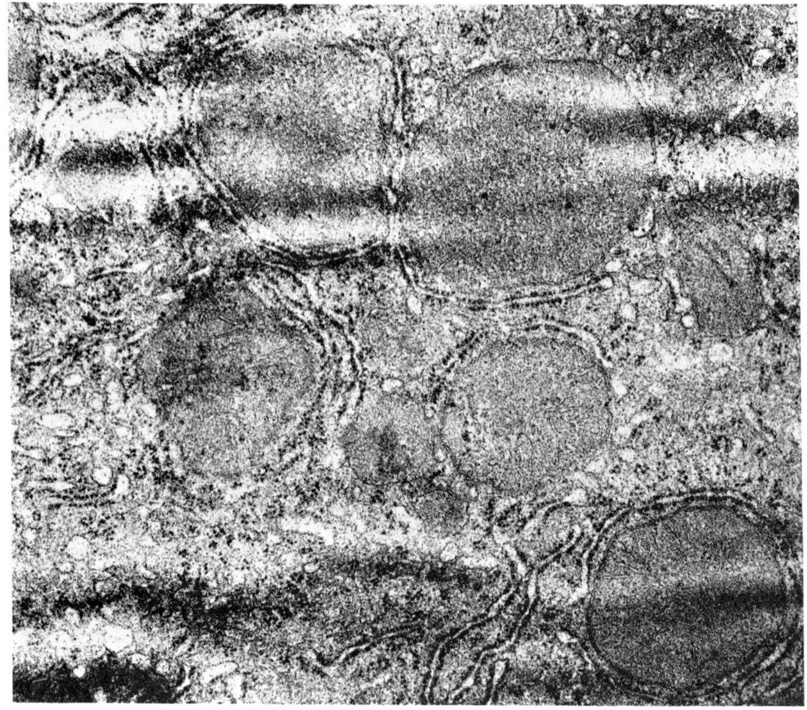

Figure 13-13. Chatter marks in an epon-embedded liver tissue cut with a glass knife.

uranyl acetate followed by lead citrate is most commonly used.

Uranyl acetate usually reacts with RNA and DNA and is a good stain for the nucleus. Lead citrate reacts with RNA, cytoplasmic membranes, and glycogen or starch in OsO_4-fixed specimens. Other stains that have been used include $KMnO_4$, uranyl nitrate, vanadium salts, ruthenium red, silver nitrate, barium and lead hydroxide, and PTA. Lead hydroxide, the most popular stain in the past, has been largely replaced by commercially available lead citrate.

A 1% solution of uranyl acetate, made either in 50% ethanol or distilled water, is commonly used. A saturated alcoholic solution is considered a better stain. Uranyl acetate dissolves slowly; a small amount must be made, and a day or two should be allowed for the solution to be dissolved. The bottle should be wrapped with aluminum foil and kept in the dark. The sediment should not be disturbed; only the clear solution should be used for staining.

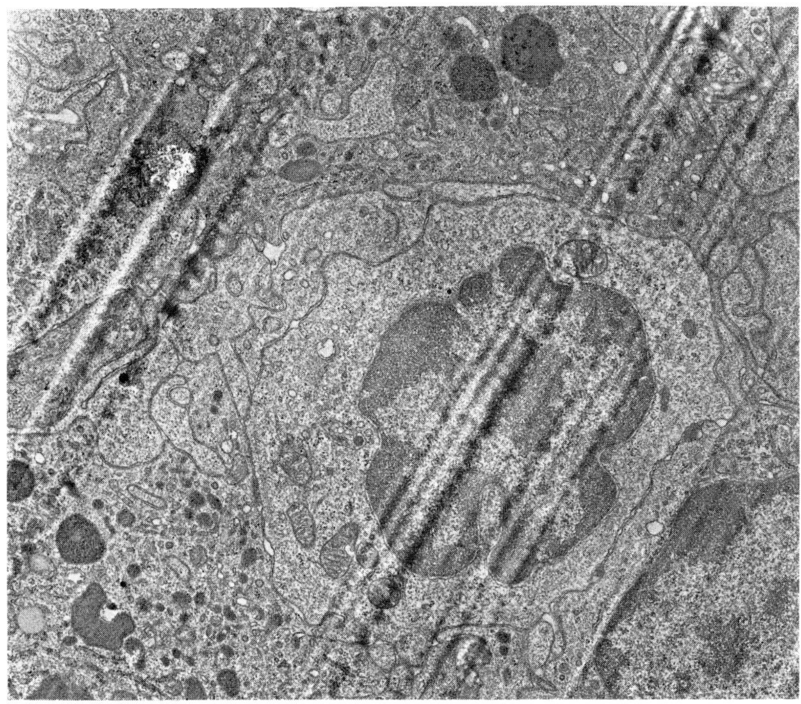

Figure 13-14. Knife marks in an epon-embedded spleen tissue cut with a diamond knife.

An alkaline lead citrate solution is made by dissolving one pellet of NaOH in 50 ml of distilled water. To this, 0.25 g of lead citrate is added, and a magnetic stirrer is used for dissolving the chemical (it dissolves slowly). Warming the solution facilitates dissolution. The stain is stable and should be stored in a tightly stoppered bottle; it matures with age. Some undissolved chemical may remain in the bottle, and the clear solution should be used for staining.

Thin section staining can be conveniently done on a sheet of parafilm. A drop of uranyl acetate is placed on this wax paper; the dried grid, sections down, is placed on the drop and covered with a petri dish. A staining time of at least 30 minutes is required. Sections can also be stained on a sheet of dental wax or a depression slide. After the uranyl acetate staining, the grid is gently washed with a jet stream of distilled water from a plastic squeeze bottle. All staining solution must be removed, otherwise a precipitate would be formed

after drying. The section is then placed on a filter paper and dried. The sections are then stained with lead citrate as described for uranyl acetate. A staining time of 2 to 5 minutes is sufficient. The grid is washed and dried afterwards. The sections are now ready for examination in the TEM. Electron micrographs of some thin sections are illustrated in Figure 13-15.

Physical Methods of Tissue Preparation

A number of attempts have been made to use physical techniques for preparing biologic specimens that do not involve chemical fixatives or dehydrating agents. These include freeze-drying, freeze-etching, directly sectioning the frozen tissue, and freeze substitution. Considerable difficulties have been encountered in preventing ice crystal formation in the frozen tissue with these techniques. Inert dehydration at room temperature has also been used.

Freeze-Drying

Very small pieces of tissues are rapidly frozen with liquid nitrogen, and the tissues are dehydrated in a vacuum in a conventional freeze-drying apparatus (lyophilizer). The tissues are then embedded, and the polymerization is allowed to proceed at a low temperature. Difficulties are encountered when infiltrating the hard tissue and in cutting. A distortion of fine structures also occurs owing to the surface tension during freeze drying.

Freeze-Etching

This technique is described in Chapter 15.

Direct Sectioning of the Frozen Tissue

The frozen tissue can be sectioned directly with a cryoultramicrotome (*see* cryoultramicrotomes in this Chap.). Electron micrographs of frozen tissue sections are shown in Figure 13-16.

Inert Dehydration and Freeze Substitution

Inert dehydration can be done at room temperature without freezing the tissue before or after glutaraldehyde fixation. Small

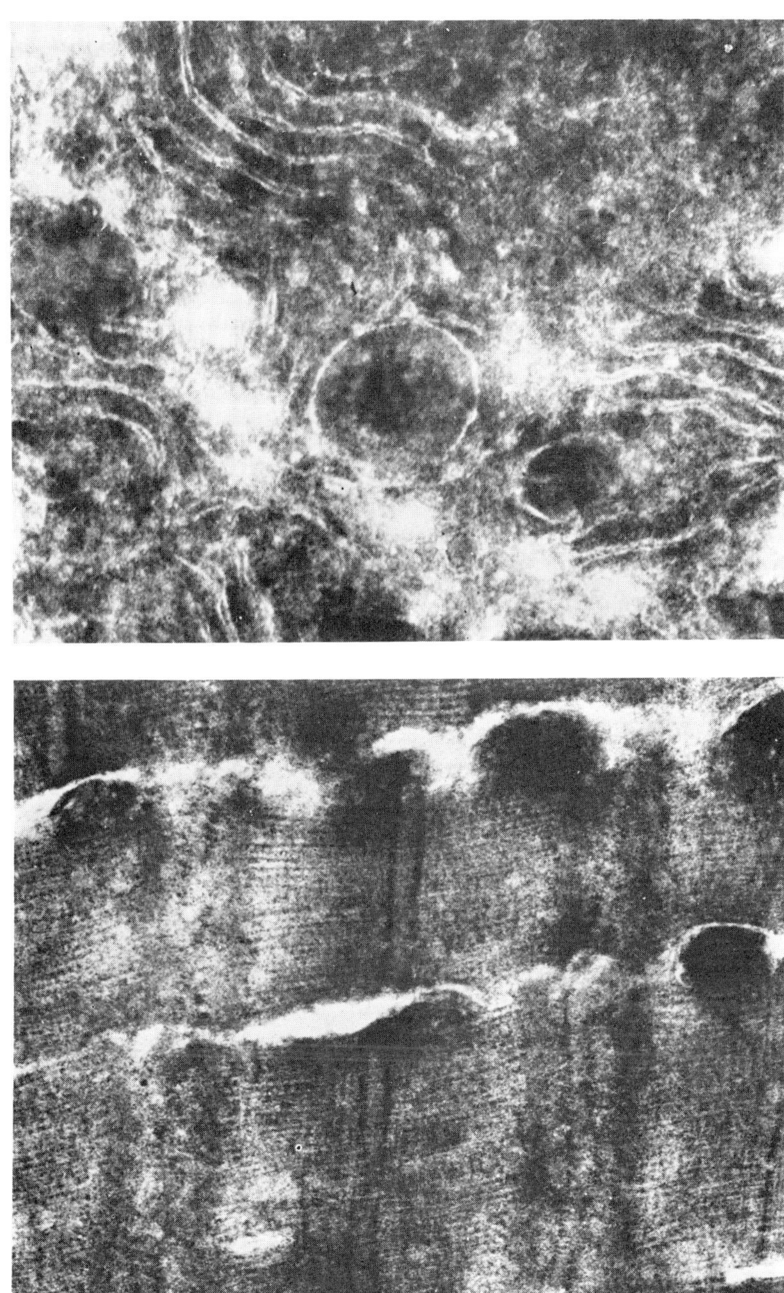

Figure 13-16. Frozen sections cut on a LKB cryoultramicrotome. (a) Unstained kidney tissue. (b) Muscle tissue stained with uranyl acetate and lead citrate. Magnification unknown. Courtesy of LKB Instruments, Inc.

pieces of tissues are passed through a graded strength of ethylene glycol (10 to 100%), and the tissues have a glassy appearance after glycol treatment. The glycol replaces water in the tissue and forms hydrogen bonding. The tissue can be fixed and block stained when it is in 66% glycol. Osmication is possible only if trimethylene glycol is used rather than ethylene glycol. The tissue is then embedded in water-soluble embedding medium, or the tissue is treated with propylene oxide before embedding in an epoxy resin (*see* in this Chap.). Inertly dehydrated specimens are not exposed to water again, and sections can be cut with ethylene glycol as the trough fluid. Uranyl acetate and PTA are soluble in ethylene glycol; the sections can be stained and the grids dried with a low vacuum. A differential shrinkage has been reported with this technique.

Freeze substitution was applied to transmission electron microscopy by Fernández-Morán in 1957. In this technique, small pieces of tissues are rapidly frozen, and the ice is replaced (substituted) by passing the tissues in a series of cryoprotective liquids such as ethylene glycol at low temperature. The tissue becomes glassy and transparent after glycol treatment, and the frozen tissue can be sectioned directly. The tissue can be frozen in Freon® 22, liquid nitrogen, or 77% glycol cooled at −77° C. The substitution is generally carried out with 70% ethylene glycol at −50 to −70° C. An overnight substitution is generally used, but sometimes the tissue may be left in the substituting medium for 1 week. The tissue is then embedded and cut as described for inert dehydration. Very few artifacts, due to secondary ice crystal formation, have been reported. However, acetone, alcohol, and OsO_4 apparently cause many ice crystal formations. Glutaraldehyde may be added to the glycol bath. Apparently there is no fundamental difference between inert dehydration and freeze substitution (Fig. 13-17).

SELECTED BIBLIOGRAPHY

Dawes, C.J.: *Biological Techniques in Electron Microscopy*. New York, Barnes & Noble, 1971.

Fawcett, D.W.: *An Atlas of Fine Structure: The Cell*. Philadelphia, Saunders, 1966.

Fernández-Morán, H.: Low temperature preparation techniques for electron microscopy of biological specimens based on rapid freezing with liquid helium. II. *Ann NY Acad Sci*, *85*: 689, 1960.

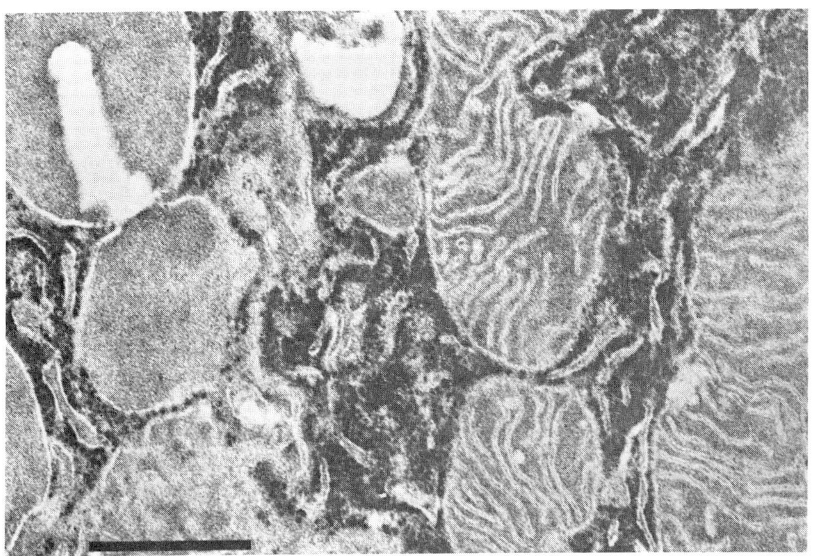

Figure 13-17. Pancreatic tissue processed by the freeze-substitution technique. Bar = 1μm. From D. C. Pease, Substitution techniques. In J. K. Koehler (Ed.), *Advanced Techniques in Biological Electron Microscopy*, 1973. Courtesy of Springer-Verlag, New York.

Glauert, A.M.: The fixation and embedding of biological specimens. In Kay, D.H. (Ed.): *Techniques for Electron Microscopy*, 2nd ed. Philadelphia, F.A. Davis, 1965.

Glauert, A.M. and Phillips, R.: The preparation of thin sections. In *Techniques for Electron Microscopy*, 2nd ed. Philadelphia, F.A. Davis, 1965.

Hayat, M.A.: *Principles and Techniques of Electron Microscopy. Biological Applications*, vol. 1. New York, Van Nostrand, 1970.

Luft, J.H.: Embedding media—old and new. In Koehler, J.K. (Ed.): *Advanced Techniques in Electron Microscopy*. New York, Springer-Verlag, 1973.

Meek, G.A.: *Practical Electron Microscopy for Biologists*, 2nd ed. London, Wiley, 1976.

Mercer, E.H. and Birbeck, M.S.C.: *Electron Microscopy—A Handbook for Biologists*. Oxford, Blackwell, 1961.

Mohanty, S.B.: A rapid embedding technique for electron microscopic study of a bovine herpesvirus. *Zentralbl Bakteriol (Orig A) 224*:61, 1973.

Pease, D.C.: *Histological Techniques for Electron Microscopy*, 2nd ed. New York, Academic Press, 1964.

Pease, D.C.: Substitution Techniques. In Koehler, J.K. (Ed.): *Advanced Techniques in Electron Microscopy*. New York, Springer-Verlag, 1973.

Rebhun, L.I.: Freeze substitution and freeze-drying. In Hayat, M.A. (Ed.): *Principles and Techniques of Electron Microscopy, Biological Applications*, vol. 2. New York, Van Nostrand, 1972.

Sjöstrand, F.S.: *Electron Microscopy of Cells and Tissues*, vol. 1. *Instrumentation and Techniques*. New York, Academic Press, 1967.

Wischnitzer, S.: *Introduction to Electron Microscopy*, 3rd ed. New York, Pergamon, 1981.

Chapter 14

COUNTING OF VIRUS PARTICLES

THE TRANSMISSION ELECTRON microscope (TEM), considered by some as a qualitative instrument, has a unique advantage in that it provides us with pictures that, without being subjected to complicated quantitative analysis, are often meaningful and pleasing. Numerous qualitative applications of this instrument have contributed significantly to the advancement of science. These include the visualization of the structural organization of cells, viruses, and the very building blocks of which living things are made, to name only a few. The electron image, however, also contains an abundance of information that can be obtained by quantitative methods. The simplest quantitative use of the TEM involving the counting of virus particles in a suspension will be described here.

Virologists generally determine the number of infective virus particles in a sample by various bioassay methods, e.g. pock or plaque titration. The direct counting of actual physical particles of virions, however, can only be done in the TEM. The number of infective virus particles determined by a biologic titration can be compared with the results obtained with the TEM. Although some noninfectious virus particles would be counted in a TEM, a reasonable agreement can often be achieved between the two methods. The general principle and the commonly used techniques for enumerating virus particles with the TEM will be discussed in this chapter.

THE PRINCIPLES AND METHODS OF PARTICLE COUNTING

The procedure for counting the particle content of a uniform virus suspension consists of counting *all* the particles from a *known* volume of suspension. Basically, virus particles from a known volume of suspension deposited on a filmed grid are counted, and the number of particles/ml is then determined. It is important that the particles be present in suitable numbers for counting, be identified clearly and individually, and be true representatives of the particles in the original suspension. The microdrop spray and sedimentation methods for counting virus particles were reported as early as 1949. A number of modifications of these basic techniques have now been devised. Four methods currently used are (a) the microdrop spray method, (b) the spreading method, (c) the sedimentation method, and (d) the pelleting and thin sectioning technique.

The first two methods incorporate readily recognizable reference particles of known concentration to the virus suspension. Although bacteria and even vaccinia virus particles have been used as indicator particles, polystyrene latex (PSL) particles are most commonly used for this purpose. The average numbers of virus particles and PSL spheres are determined, and from their ratio and the known concentration of PSL particles, the number of virus particles in the original suspension is calculated. The concentration of PSL particles is calculated by the following formula:

$$\text{No. of PSL particles/ml} = \frac{\text{Dry weight of 1 ml of PSL spheres}}{\text{Weight of one PSL sphere (= volume} \times \text{density)}}$$

$$= \frac{W}{{}^1\!/_6 \times \pi \times d^3 \times \rho}$$

where W = dry weight of 1 ml of PSL spheres; d = average diameter of PSL particles (supplied by the manufacturer); and ρ = density of PSL suspension (given by the supplier or can be determined, usually ≈ 1.05 g/ml).

The average diameter of PSL spheres can be checked by comparing with a reference standard, e.g. a diffraction grating replica. It is emphasized that the use of aggregated PSL spheres must be avoided if maximal accuracy is desired. Most workers have used a PSL concentration of 10^{10} particles/ml. The virus concentration is then calculated by the following equation:

No. of virus particles/ml =

$$\frac{\text{No. of virus particles in a given area}}{\text{No. of PSL spheres in the same area}} \times \begin{array}{l}\text{Actual}\\ \text{concentration of}\\ \text{PSL particles/ml}\end{array}$$

The Microdrop Spray Method

This method was developed by Williams and Backus (1949) for counting bushy stunt virus particles. In this method, the virus suspension is mixed with a known concentration of PSL spheres that usually yields a constant ratio of the two particles in a randomly selected admixed sample. A small volume of the mixture is sprayed with a high velocity spray gun so that minute droplets (2 to 20 μm dia) are deposited on a filmed grid. By using shadow casting for enhancing the contrast of their specimen and by adding traces of bovine serum albumin to their mixture, Williams and Backus were able to observe clearly the whole area on which the drop had dried. They counted all the particles of each kind in each droplet and established their ratio. Since the actual concentration of the reference PSL particles is known, the number of virus particles/ml is calculated. A rather high concentration of virus and PSL particles ($\approx 10^9$ to 10^{10} particles/ml) is required for this method because of the small sample size in the droplet. The expected error of the virus count by this method varies from 15 to 30 percent. The authors were able to calculate the molecular weight of the virus by combining the dry weight of a known volume of virus suspension with the particle count/ml.

Spreading Methods

A variation of the microdrop spray technique, these methods also employ reference particles, but they have the advantage of simplicity. In these procedures, the virus and PSL spheres are counted over a substantial number of areas, and their ratio is established. The number of virus particles/ml is then calculated from the actual concentration of latex particles. A number of these techniques have been described.

Agar Filtration Method

Kellenberger and Arber (1957) used this technique. A virus suspension is mixed with a known concentration of PSL spheres, and the mixture is spread with a smooth glass rod on collodion-coated agar plates. Collodion acts as a dialyzing membrane for the fluid and salts in the suspension, and uniformly dispersed particles are deposited on the collodion film. The preparation is then fixed in 40% formalin vapor for seven minutes, the film floated onto 2% lanthanum nitrate (used for facilitating the release of the film), shadowed, and then counted in the TEM. The concentration of virus particles/ml is calculated by the equation given previously. The accuracy of the virus count is ±20 to 30 percent.

Lowered Drop Method

For this method (Pinteric and Taylor, 1962), a small drop (2 to 4×10^{-3} ml) of a mixture of virus and PSL spheres is applied onto a small area on a formvar film floating on an ammonium acetate buffer. After thirty minutes, the film is lowered on submerged silver or gold grids (copper grids may react with the acetate buffer). The contrast of the preparation can be enhanced either by shadowing or negative staining with phosphotungstic acid (PTA), and the particles are counted in the microscope. A virus particle concentration of 5×10^7 particles/ml can apparently be counted by this technique.

Loop Drop (Single Drop) Method

This simplest method of enumerating virus particles was reported by Watson (1962). The virus-PSL particle mixture is directly applied onto a coated grid with a welded platinum loop. A 0.5% solution of PTA and 0.5% bovine serum albumin are also added to the mixture, and the negatively stained virus particles are counted by recognizing their characteristic ultrastructure. Alternatively, a *single drop* of virus-PSL particle mixture can also be applied onto the grid with a Pasteur pipette (Fig. 14-1). Several variations of the loop drop technique have subsequently been reported. A drop of virus-PSL sphere mixture can be applied onto a filmed grid floating on water that had been pretreated with 0.05% bovine serum albumin (Geister and Peters, 1963). A successful counting of samples containing as few as 10^5 particles/ml has been reported by this technique. The concentration of virus particles is calculated as described previously.

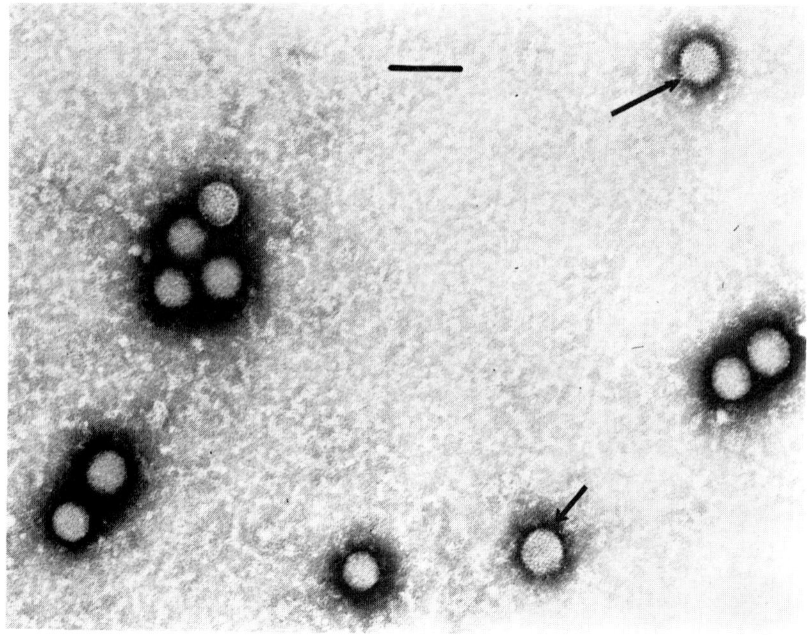

Figure 14-1. Bovine adenovirus type 3 (arrows) and latex particles counted by the single-drop spreading method. Negative stain with phosphotungstic acid. Bar = 100 nm.

Counting of virus particles from unpurified virus preparation by the negative staining technique using vaccinia virus as reference particles has also been reported (Smith, 1964).

Sedimentation Methods

Particle counting by the sedimentation methods is usually somewhat more complicated and time-consuming than the methods previously described. However, they provide a greater sensitivity, and suspensions of low concentration (10^6 particles/ml) can be counted. Soluble, nonsedimentable contaminating materials that are usually deposited on the support film can also be avoided by this procedure. The virus preparations from nonconcentrated lysates of infected cells can be counted. Unlike the microdrop spray or spreading methods, no reference particles are required for virus concentration calculation.

The basic method was developed by Sharp in 1949. Particles from a known volume of virus suspension are sedimented directly onto a receiving surface in a high-speed centrifuge. The virus particles are then pseudoreplicated (*see* Chap. 12), shadowed or negatively stained, and counted in a TEM (Fig. 14-2). A low enough magnification that enables the virions to be readily recognized should be used. The virus concentration is calculated by the following formula:

$$N = \frac{n\,M^2\,D}{0.91\,A\,h}$$

where N = no. of virus particles/ml; n = average number of virus particles counted in the known photographed field; A = area of the photographic negative (cm^2); h = height or thickness of the liquid layer that contributed its particles to the area A; M = magnification; and D = dilution factor. The factor 0.91 is used for correcting the taper of the centrifuge cell wall. The precision of the method has been calculated to be ±16 percent of the mean.

Originally collodion-coated glass coverslips were used for sedimenting the virus particles. Agar blocks were then used, and later

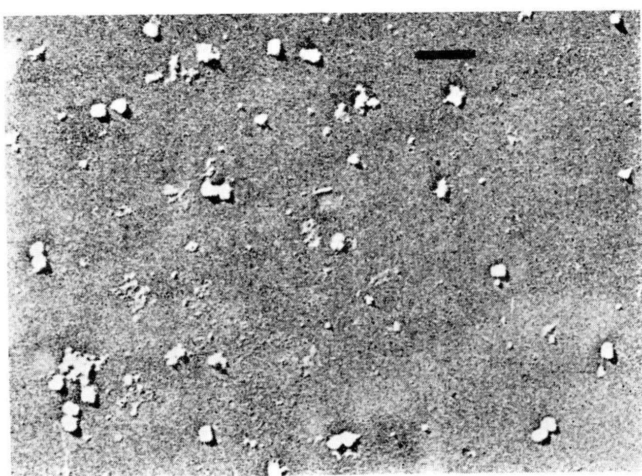

Figure 14-2. Shadowed vaccinia virus particles counted by the sedimentation technique. Bar = 1 μm. From D. G. Sharp, Quantitative use of the electron microscope in virus research. Methods and results of particle counting. In G. F. Bahr and E. H. Zeitler (Eds.), *Quantitative Electron Microscopy*. Copyright © 1965, The Williams and Wilkins Co., Baltimore.

Sharp and Overman (1958) designed a special rotor containing eight cells that is available from Sorvall-DuPont, Inc. Agar blocks, ≈2 mm thick × 1 cm^2 are placed in the cell, and a known volume of virus suspension is centrifuged to sediment the particles. The particles are then pseudoreplicated with collodion or formvar, shadowed or stained, and counted. A filmed grid has also been placed in the cell for sedimenting the particles. The centrifugation time and speed are chosen according to the size of the virus particle.

The original sedimentation method has been modified to suit variable experimental conditions. Various centrifuge cells have also been designed for the Beckman ultracentrifuge, swinging bucket type of rotor, and for virus particles that are purified by rate zonal centrifugation in a density gradient.

Pelleting and Thin Sectioning

The previously described procedures of counting virus particles by shadowing or by negative staining usually require a reasonably pure virus preparation so that the virions can be recognized while counting. Some virus particles, e.g. budding viruses, *C* type of oncoviruses, are usually pleomorphic and may not be recognized by these techniques. They often show characteristic morphology only in thin sections. A quantitative estimate of viral purity is also possible by this technique.

In these methods, the virus-infected cells are pelleted, fixed, and embedded in special bottleneck Beem capsules (Gehle and Smith, 1970). The pellets from the bottleneck are then cut and flat embedded after rotating 90 degrees. The number of virus particles in sections of known volumes (Fig. 14-3), the pellet height and diameter are determined, and the number of virus particles/unit volume is then calculated. A standard deviation range of 2 to 13 percent of the mean has been reported for these methods.

A variation of this technique was described by Miller et al. (1973). Virus particles from a cell suspension are sedimented onto Millipore® filter disks (4 mm dia). These are then embedded, thin sectioned, virus particles in these sections are counted, and their number per unit volume is estimated.

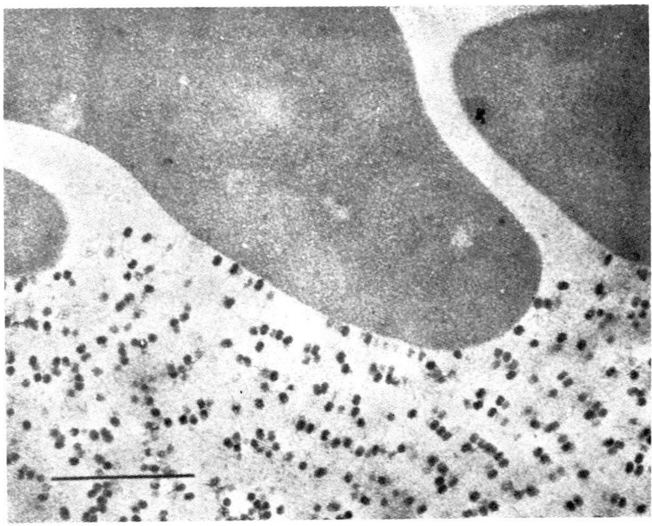

Figure 14-3. T_4 phage particles capped with sheep erythrocytes, pelleted and thin sectioned for counting. Bar = 1 μm. From W. D. Gehle and K. O. Smith, Enumeration of virus particles in ultrathin sectioned pellets, *Proceedings of the Society for Experimental Biology and Medicine, 135*:490, 1970. Courtesy of publisher.

Sensitivity of Different Counting Methods

Different workers differ in their statements of accuracy of their counting methods. This discrepancy may partly be due to the willingness of these workers to count or scan large areas. To avoid these variables, it should be realized that regardless of the method used, the particles seen per unit area are directly proportional to the quantity of the fluid from which they came. Sensitivity is inversely proportional to the square of the magnification, and a greater accuracy is obtained by using the lowest microscope magnification, provided the virus particles can be recognized at this magnification.

Problems Encountered in Particle Counting

The recognition of virus particles among cellular debris, especially in crude virus-infected lysates, may be most troublesome while counting the particles. This should not be difficult, however, if the preparation is relatively pure and the particles have a well-defined shape. If the fine structure (by negative staining) is used as a means for identification, there should be no problem in deciding what to count.

Difficulties may also arise in preservation of virus particles, particularly those which are extremely fragile, e.g. herpes viruses. The negative staining technique may be used for these agents. Since the number of particles counted has to be related to the volume in which they were originally suspended, it is important that the particle distribution be uniform in the suspension. Some PSL particle preparations may be badly clumped, and if used, they may greatly influence the accuracy of the count.

Choosing a Method

Virtually all methods described here have their limitations, and the choice of a particular counting technique would depend on the need of the microscopist. The sedimentation method usually offers a greater accuracy than any other methods, but it needs special equipment and is time-consuming. Ultracentrifuges are standard equipment in almost all laboratories, and either by purchasing or fabricating his own centrifuge cells, the microscopist could use this method. Some sedimented particles, however, may not firmly adhere to the receiving surface and may be lost during pseudoreplication. The level of cell debris may also be a limiting factor, but an enzyme treatment may be used to reduce the debris. Pelleting and thin sectioning, a modified sedimentation technique, is probably the method of choice for budding oncoviruses.

The spreading method using reference PSL spheres, especially the single drop method, is undoubtedly the easiest, quickest, and least expensive method for counting virus particles. The preparation can be negatively stained for counting only those particles which have characteristic ultrastructures. However, the particles tend to be concentrated near the edge of the grid as the drop dries on the support film. Therefore, large areas have to be scanned for ascertaining the representative ratio of virus particles to PSL spheres.

Once the microscopist chooses a particular technique, it is advisable to check the accuracy of the results with those obtained with another technique. A linear relationship can also be obtained for several dilutions of a given suspension.

SELECTED BIBLIOGRAPHY

Gehle, W.D. and Smith, K.O.: Enumeration of virus particles in ultrathin sectioned pellets. *Proc Soc Exp Biol Med, 135*:490, 1970.

Geister, R. and Peters, D.: Ein Vereinfachtes direktes Zahlverfahren für Virus-Suspension ab 10^5 Partikel/ml. *Z Naturforsch, 18b*:266, 1963.

Kellenberger, E. and Arber, W.: Electron microscopical studies of phage multiplication. I. A method for quantitative analysis of particle suspensions. *Virology, 3*:245, 1957.

Mathews, J. and Buthala, D.A.: Centrifugal sedimentation of virus particles for electron microscopic counting. *J Virol, 5*:598, 1970.

Miller, M.F., Allen, P.T., and Dmochowski, L.: Quantitative studies of oncornaviruses in thin sections. *J Gen Virol, 21*:57, 1973.

Miller, M.F. II: Particle counting of viruses. In Hayat, M.A. (Ed.): *Principles and Techniques of Electron Microscopy: Biological Applications*, vol. 4. New York, Van Nostrand, 1974.

Pinteric, L. and Taylor, J.: The lowered drop method for preparation of specimens of partially purified virus lysates for quantitative electron micrographic analysis. *Virology, 18*:359, 1962.

Sharp. D.G.: Enumeration of virus particles by electron micrography. *Proc Soc Exp Biol Med, 70*:54, 1949.

Sharp, D.G. and Overman, J.R.: Enumeration of virus particles in crude extracts of infected tissues by electron microscopy. *Proc Soc Exp Biol Med, 99*:409, 1958.

Sharp, D.G.: Quantitative use of the electron microscope in virus research. Methods and recent results of particle counting. In Bahr, G.F. and Zeitler, H. (Ed.): *Quantitative Electron Microscopy*. Baltimore, Williams and Wilkins, 1965.

Smith, K.O.: Methods of staining and counting virus particles from unpurified materials. *Bact Proc, 64*: 164, 1964.

Strohmaier, K.: A new procedure for quantitative measurements of virus particles in crude preparations. *J Virol, 1*:1074, 1967.

Watson, D.H.: Electron-micrographic particle counts of phosphotungstate-sprayed virus. *Biochim Biophys Acta, 61*:321, 1962.

Williams, R.C. and Backus, R.C.: Macromolecular weights determined by direct particle counting. I. The weight of the bushy stunt virus particle. *J Am Chem Soc, 71*:4052, 1949.

The treated specimen is usually placed in a copper specimen holder and dipped in liquid Freon 22 cooled close to its freezing point (−165° C) with liquid nitrogen. The specimen can then be transferred to the freeze-etching apparatus for further processing or stored in liquid nitrogen (several weeks) for later use. The basic freezing step has been modified by many workers depending on their needs and the instrumentation available.

Fracturing

A cutting or *fracturing* of the frozen specimen exposes the internal details, and the *freeze-fracturing* appears to be a brittle fracturing. The fracturing produces a series of roughly hemispherical breaks in front of the knife edge. The fracture also appears to occur along the interior planes of membranes, probably because of their plane of weakness. The hydrophobic interior portion of the membrane is usually exposed by the fracture process.

Various freeze-etching devices use different methods to accomplish fracturing, and this technique appears to be undergoing constant modification. In the microtome freeze-etchers (Moor-Blazer unit), a remote control mechanically advanced cutting edge is used for fracturing. A thermal advance mechanism for this purpose is also available. For the Steere-Denton freeze-etching apparatus, a long handled scalpel manipulated from outside was originally used for this purpose. The unit has now been modified (Fig. 15-1) so that the frozen specimen held in a hinged complementary replica holder is forced apart in the vacuum chamber to cause the fracture. A preset spring-loaded trigger mechanism releases a stainless steel knife for one cutting in the McAlear-Kreutziger block module marketed by Ebtec Corporation. For the Bullivant and Ames Type II block freeze-etching unit, the frozen specimen is fractured with a razor blade precooled with liquid nitrogen before transferring the block apparatus to the vacuum chamber.

Apparently cutting with a knife to cause the fracture is not required, and many workers prefer to place the tissue in a hinged complementary replica holder that can be forced apart for fracturing the frozen tissue. Freeze-fracture with little or no etching is used for many studies.

Figure 15-1. Steere-Denton freeze-etch unit. Courtesy of Dr. R. L. Steere.

Etching (Ice Sublimation)

When some ice from the fractured surface of the tissue is *sublimed* (*etched*) away, intra- or extracellular materials that are otherwise masked from view are exposed that yield additional information of internal details. For some specimens, the etching step is apparently not required, and indeed, it may obscure the details present in the original fractured specimen.

After fracturing, the specimen is *freeze-etched* by keeping it at $-100°$ C near a cold trap before replication. The vapor pressure of ice at $-100°$ C is 1×10^{-5} torr, which gives an etching *rate* of ≈ 20 Å / second. The *etching time* may vary depending on the structures desired to be seen, but an etching time of thirty seconds to one minute that sublimes ≈ 600 to 1,200 Å of ice is usually adequate for most specimens. It is critical that the temperature be maintained at $-100°$ C $\pm$ 1° C, otherwise the magnitude of the etching rate could be severely altered. The hydrophilic portion of the membrane (true surface) is usually exposed by the etching process.

It is also important to prevent the contamination that is deposited on the cold, exposed fractured surface of the specimen during ice sublimation. The liquid nitrogen-cooled knife holder is left in position over the specimen until just before replication in the microtome type of freeze-etching devices. An outer shroud surrounds the specimen to prevent contamination in the Steere-Denton freeze-etcher. In the block type of devices, this is achieved by the relatively long, cold pathway of most contaminants through the upper block trap before they are deposited on the specimen.

Replication

The fractured or etched cold specimen surface is usually replicated by depositing 20 to 50 Å of platinum-carbon at an angle of 45° by the resistance heating method (*see* Chap. 11). This evaporant has a grain size of ≈ 20 Å, and this is the limit of the specimen detail that can be resolved in the replica. A thin carbon film is evaporated normal to the specimen surface after platinum-carbon shadowing for strengthening the replica. Some workers have reported good results by shadowing in an electron beam evaporator (*see* Chap. 11).

Cleaning of Replica

This involves a chemical removal of the tissue so that the replica would be released, washed, picked up on the grid, and examined in the TEM. It is emphasized that platinum-carbon replicas are extremely brittle, and they should be carefully handled. The replica end of the tissue may be dipped in 1% collodion solution in amyl acetate for helping it to hold together during subsequent steps.

The specimen is removed from the freeze-etching apparatus and is allowed to thaw by placing it in distilled water or in the original solution used for pretreatment before freezing. The cleaning solution then replaces the water or suspending medium. Small glass petri dishes, depression slides, or depression procelain plates can be used for this purpose. Replicas of cell pellets are generally released easily, but tissue replicas often tightly adhere to the underlying tissue. A number of cleaning solutions have been used for digesting the tissue, but the ordinary household chlorine bleach (5% sodium hypochlorite) has been most successful especially for animal tissues. However, this agent generates some bubbles that may cause some difficulties and disruption of the replica. Other cleaning solutions used include chrome-sulphuric acid (most suitable for plant tissues), household bleach followed by 70% H_2SO_4, hot HNO_3, strong

NaOH, a combination of all these, and proteolytic enzymes. The length of time is usually determined empirically (generally 30 minutes).

After cleaning, the replicas are carefully rinsed in several changes of distilled water. They can be carried with a platinum loop and placed on the surface of fresh water. Otherwise, the cleaning solution can be carefully removed with a capillary pipette and replaced with distilled water a number of times. The replica is then picked up on an uncoated 200- or 300-mesh grid or on a grid with a formvar net.

Instrumentation

Three basic types of freeze-etching instruments (discussed previously) are available, and it is not possible to describe them in detail here. Some of them are quite elaborate and expensive, and some give more reliable results than others. The Blazer microtome type of machine is the most expensive, and the Type II block type of freeze-fracture unit (Ebtec Corp.) is the least expensive.

Complementary Replicas

Normally, only one-half of the fractured specimen is replicated (Fig. 15-2) by the freeze-fracture or freeze-etching technique, and the other half of the specimen is usually chipped away during fracturing the frozen specimen. However, it is evident that it would be extremely useful to examine the replicas from both the faces of the tissue produced by a single fracture. Thus, the complementary regions, e.g. the number of particles and corresponding pits, could then be matched. Various techniques, similar in approach, have been used to obtain double-replica freeze-fractured or freeze-etched preparations. Essentially, the specimen is placed in a specimen holder that can be broken into two matching halves. The two corresponding fractured or etched halves thus exposed are then replicated with platinum-carbon in the same plane, and the replicas examined (Fig. 15-3).

In the Steere's device, the specimen is placed in a small groove of a hinged device made of solid gold, which can be forced apart in the

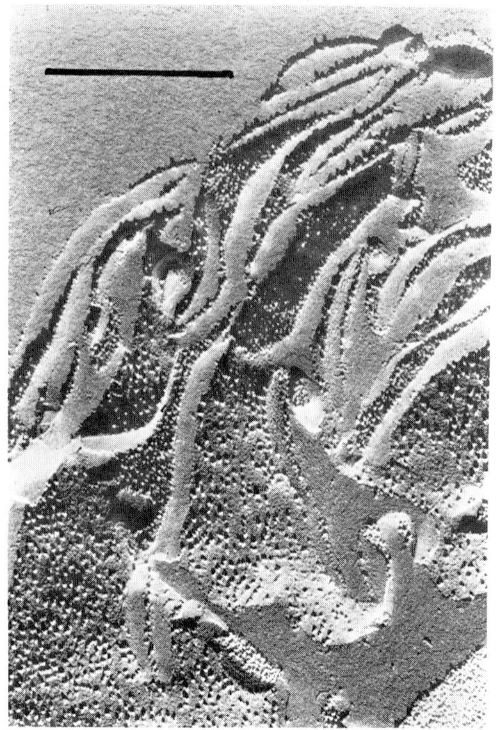

Figure 15-2. Freeze-fractured and freeze-etched preparations. (a) Freeze-fractured chloroplast. Bar = 0.5 μm.

vacuum chamber (Fig. 15-4). A modified double replica device can be used to prepare as many as ten double replicas.

Interpretation

Interpretation of freeze-etched images of the cellular components may be quite difficult. Artifacts are generally introduced during preparation of all biologic specimens for electron microscopy, and this can be very disturbing with freeze-fracture or freeze-etch preparations. Intracellular ice crystal formation varies from being barely perceptible to those that are massive and almost fill the entire cell. Large ice crystals generally have an angular shape and are easily recognized, and they usually indicate the inadequacy of the cryoprotecting treatment. Ice crystal formation is barely visible in

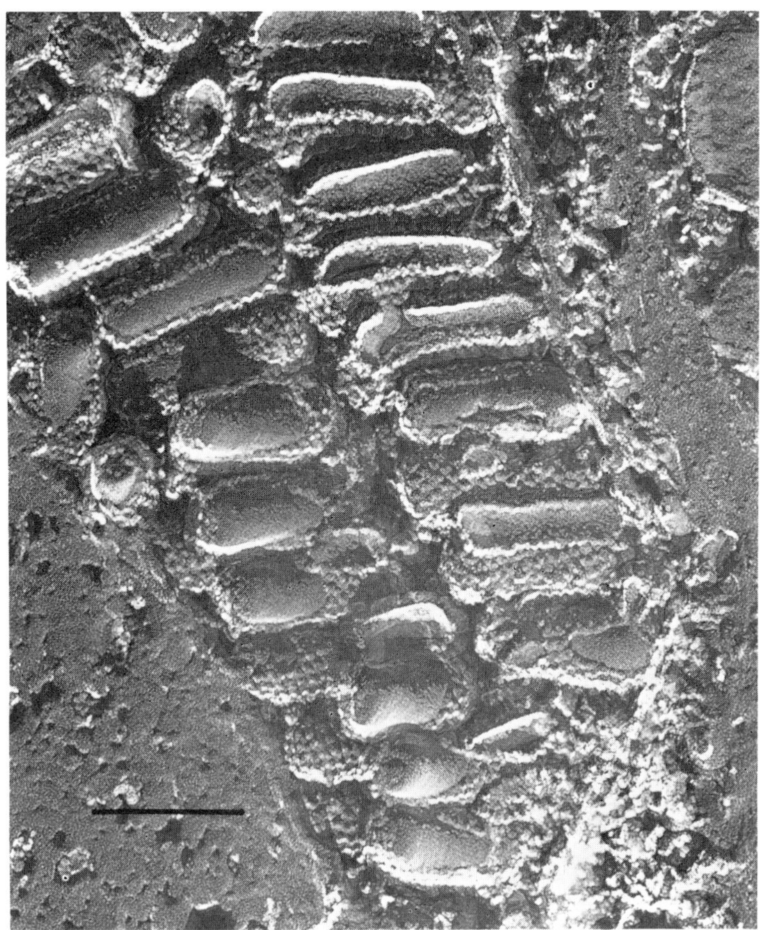

Figure 15-2. (b) Freeze-etched potato yellow dwarf virus in the nucleus of a *Nicotiana rustica* cell showing virus particles and their structures. Bar = 100 nm. Both courtesy of Dr. R. L. Steere.

freeze-fracture specimens, but this is usually revealed after etching.

The process of fracturing may also cause problems in interpretation. The best ultrastructural information is obtained from areas that have not been actually cut by the knife but rather where the fracture occurred along a preferential weak plane. This is why many workers do not use the knife for causing the fracture. A deformation of organelle structures into abnormal configurations may be caused during the fracturing process. Hydrocarbon contamination in the vacuum evaporator can also introduce artifacts.

Relating the replica images to cellular organelles is probably the most difficult task. The orientation of the micrograph appears to be important for providing a qualitative visual impression of depth of

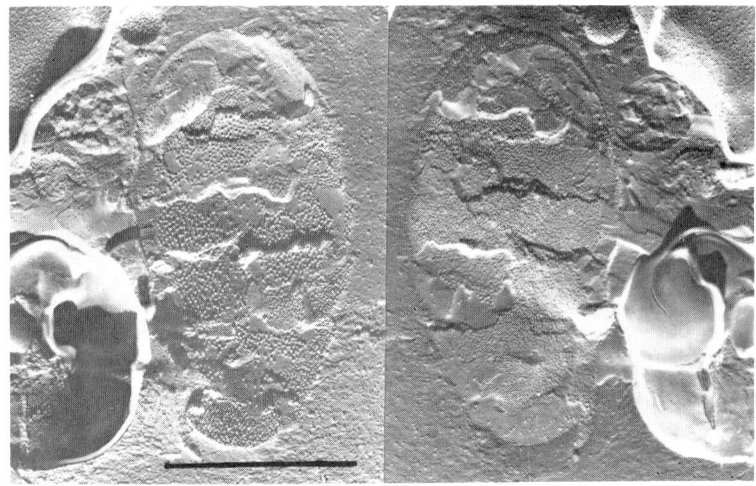

Figure 15-3. Complementary replica of freeze-fractured chloroplast. Bar = 1 μm. Courtesy of Dr. R. L. Steere.

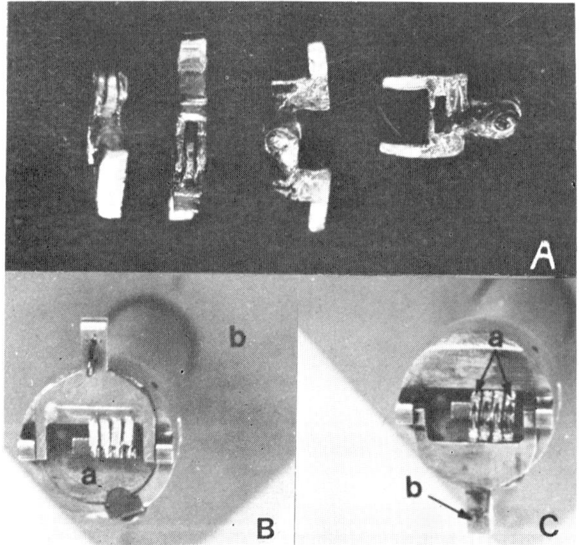

Figure 15-4. Complementary replica holder for Steere-Denton freeze-etch unit. (A) Hinged gold specimen holder in open and closed configuration. (B) The holder in closed position before fracture in the specimen holder cap. (C) Specimen holder after fracture in the unit. (a) Open specimen holder that has been forced open by the fracture arm of the cap (b). All courtesy of Dr. R. L. Steere.

the shadowed preparation. A stereo pair (Fig. 15-5) electron micrograph gives additional information for three-dimensional reconstruction.

Complementary replicas, surface labelling, and thin sections of freeze fracture tissues indicate that the membranes split along a unique interior plane. Extensive views of the specimen membranes can be obtained by freeze-etching techniques. After the membranes split, two new fracture faces from the hydrophobic interior of the membrane are seen; the true surfaces can be exposed by etching. For any membrane that can be split, the half closest to the cytoplasm,

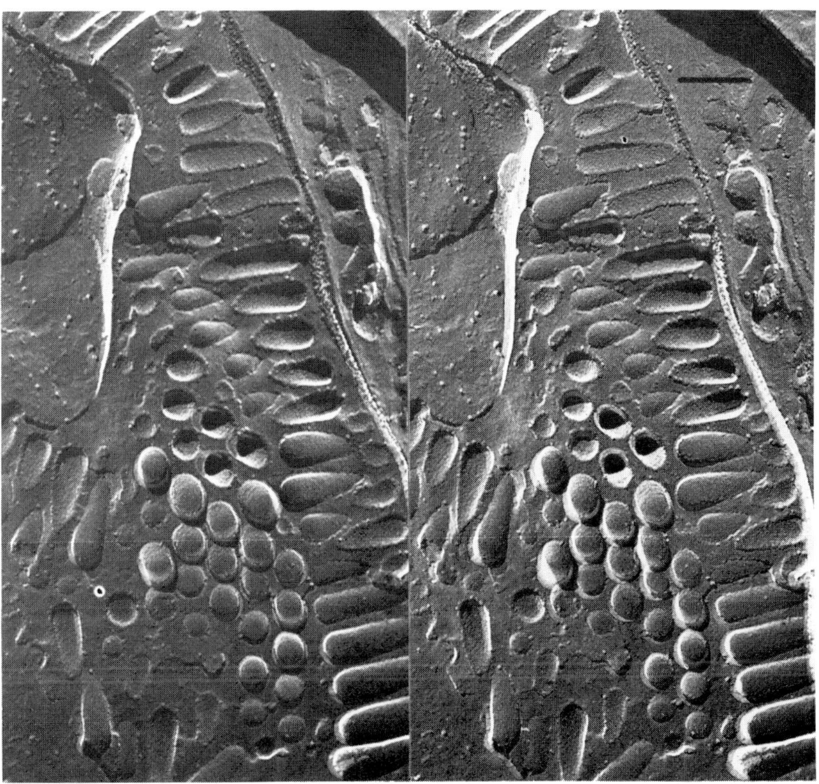

Figure 15-5. A stereo pair of potato yellow dwarf virus particles in the nucleus of a *Nicotiana rustica* cell. Freeze-fractured. Bar = 100 nm. Courtesy of Dr. R. L. Steere.

nucleoplasm, chloroplast stroma, or mitochondrial matrix is commonly designated the *protoplasmic half* (P). The half closest to the extracellular space, exoplasmic space, or endoplasmic space is designated *extracellular, exoplasmic, or endoplasmic half* (E). The hydrophobic portion of the membrane exposed after the fracture process is called PF or EF; the hydrophilic portion of the membrane exposed after etching is called PS or ES. A single membrane, then, will contain two surfaces, PS and ES, and two fracture faces, PF and EF. Since little or no etching is done for many studies, simply P and E can be used to label the micrographs.

Particles observed on virtually all membrane faces are apparently integral parts of the membrane. However, some small particles not firmly held by ice may be removed during the sublimation process. The rough and smooth endoplasmic reticulum are also very difficult to distinguish, and a deep etching may be needed to demonstrate the ribosomes.

Finally, it is emphasized that freeze-fracture and freeze-etch techniques are adjuncts to the conventional ultrathin sectioning method. Together, these techniques can provide extremely valuable information that cannot be obtained with either of the techniques alone. Although freeze-fracture and freeze-etch techniques have been developed to a point where they can now be considered routine techniques, they still require considerable skill. A microscopist new to the field does not attain early success with these methods. Before starting on his own, he should spend some time in a laboratory where the techniques are well established.

SELECTED BIBLIOGRAPHY

Branton, D., Bullivant, S., Gilula, N.B., Karnovsky, M.J., Moor, H., Muhlethaler, K., Northcote, D.H., Packer, L., Satir, B., Speth, V., Stalhlin, L.A., Steere, R.L., and Weinstein, R.S.: Freeze-etching nomenclature. *Science, 190*:54, 1975.

Bullivant, S.: Freeze-etching and freeze-fracturing. In Koehler, J.K. (Ed.): *Advanced Techniques in Electron Microscopy*. New York, Springer-Verlag, 1973.

Koehler, K.: The freeze-etching technique. In Hayat, M.A. (Ed.): *Principles and Techniques of Electron Microscopy. Biological Applications*, vol. 2. New York, Van Nostrand, 1972.

Miller, K.R.: Freeze-etching studies of the photosynthetic membrane. In Griffith, J.D. (Ed.): *Electron Microscopy in Biology*, vol. 1. New York, Wiley, 1981.

Steere, R.L.: Electron microscopy of structural detail in frozen biological specimens. *J Biophys Biochem Cytol*, *3*:45, 1957.

Steere, R.L., Erbe, E.F., and Moseley, J.M.: Prefracture and cold-fracture images of yeast plasma membrane. *J Cell Biol 86*:113, 1980.

Steere, R.L.: Preparation of freeze-fracture, freeze-etch, freeze-dry, and frozen surface replica specimens for electron microscopy in the Denton DFE-2 and DFE-3 freeze-etch units. In Johnson, J.E., Jr. (Ed.): *Current Trends in Morphological Techniques*, vol. 2. Boca Raton, CRC Press, 1981.

Chapter 16

CYTOCHEMICAL, IMMUNOLOGIC, AND OTHER SPECIALIZED TECHNIQUES

TRANSMISSION ELECTRON microscopy has now been combined with various biochemical and cytochemical techniques to provide an insight into the functional significance of cellular ultrastructures. The application of the transmission electron microscope (TEM) has also been extended to immunocytochemical and other specialized studies. These methods are undergoing constant modifications and refinement. The commonly used techniques are described in this chapter.

LOCALIZATION OF ENZYMES (ENZYME CYTOCHEMISTRY)

Cytochemical staining methods for localization of enzymes in thin sections by transmission electron microscopy are more complex than light microscopy, and modifications are necessary for each enzyme studied. The basic principle of the technique is to deposit small electron-dense particles in the tissue as the end product of a controlled chemical reaction and correlate the ultrastructure with the function. In these techniques, it is important that the enzymatic activity and cellular ultrastructure be maintained during preparation and staining of the specimen. The end product deposited at active sites must also have sufficient contrast in relation to the surrounding structure without obscuring the morphology. Therefore, a delicate choice of correct conditions is required for meaningful localization of enzymic activity and faithful preservation of ultrastructure.

Basically, the tissue is fixed in an aldehyde medium; the fixed tissue is usually frozen and cut into thick sections (10 - 50 μm) on a cryostat. It is then incubated and stained in the cytochemical medium containing the substrate, rinsed, postfixed in OsO_4, dehydrated, and thin sectioned for electron microscopy. Fixation with OsO_4 drastically reduces the enzymatic activity and is unsuitable for this study. Glutaraldehyde preserves high enzyme activities and the fine structures and has replaced formaldehyde to a large extent. Some workers have added H_2O_2 to glutaraldehyde for better preservation of fine structures as well as enzymes. Cryoprotective agents such as dimethyl sulfoxide, glycerol, and glycols have also been used to treat the tissue prior to glutaraldehyde fixation.

Only purified glutaraldehyde should be used for enzyme localization because commercial glutaraldehyde has many impurities such as ethanol, glutaric acid, and acrolein that inactivate enzymes. Glutaraldehyde can be purified by treating it with activated charcoal followed by vacuum distillation. The oily, clear solution has a maximum absorption at 280 nm. Glutaraldehyde can also be purified on a Sephadex G-10 column. Purified glutaraldehyde in sealed ampoules is now available, and it should be stored at 4° C. Fixation reduces diffusion of enzymes and increases binding of dyes to proteins, thus insuring a better localization of the remaining activity. Blocks of aldehyde-fixed nonfrozen tissues apparently cause nonuniform staining of the outer layer of cells. The optimal incubation time in the cytochemical medium should be empirically established. An incubation time of 5 minutes to 1 hour has been used, but overincubation should be avoided.

Various types of cytochemical staining reactions for different enzymes have been developed. These include (a) certain *metallic salts* that can be precipitated as a result of enzymic activity (Fig. 16-1). They are excellent because their density is very high, and they are insoluble in dehydrating and embedding media. However, they are deposited heavily, and this may obscure fine structures, and some metallic ions may inhibit the enzymic activity. The majority of enzyme localization studies have employed the formation of insoluble lead salts, usually lead phosphate, or lead sulphide. Lead sulphide has been deposited in tissues by hydrolysis of thiolacetate in the presence of lead ions to demonstrate the sites of esterase activity. Hydrolysis of potassium tellurite in the presence of lead ions also has been used to detect the site of succinic dehydrogenase

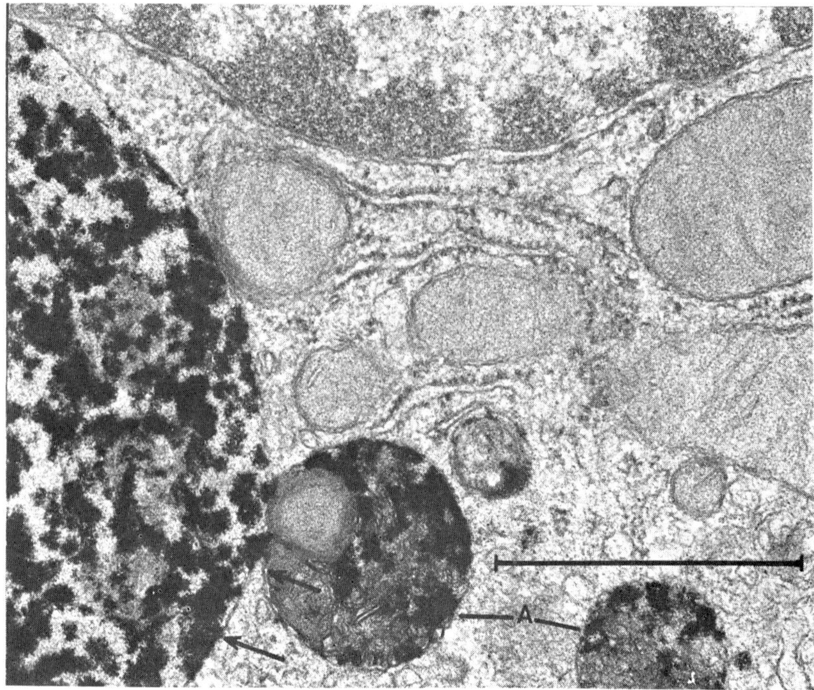

Figure 16-1. Portion of a hepatocyte incubated for acid phosphatase activity by the lead procedure in cytidine 5′ monophosphate medium for 30 minutes at 37° C. Reaction product is seen in the protein droplet (arrow) at the left and in the autophagic vacuole. The body indicated by A, at the right, is either an autophagic vacuole or residual body. Bar = 1 μm. Courtesy of Dr. A. B. Novikoff.

activity in the cell. The Gomori acid phosphatase lead medium contains β-glycerophosphate as substrate and lead nitrate. Cytidine 5′-monophosphate has substituted β-glycerophosphate in some modified Gomori media. Acid phosphatase activity has been demonstrated in various types of lysosomes, endoplasmic reticulum, and Golgi bodies by this technique.

(b) *Azo dyes* have been successfully applied to form nonmetallic, amorphous, organic compounds of enzymatic products. Unlike the metallic salts, azo dyes appear to undergo diffusion during dehydration and embedding, and their density is low. Azo dyes such as naphthol AS acetate and naphthol AS phosphate have been used for this purpose.

(c) Some enzyme substrates contain reacting groups that selectively bind with OsO_4 after osmication, giving rise to deposition of *osmium black* at the reaction sites. These sites are apparently not affected during dehydration and embedding. Peroxidase has been localized by using 3,3′-diamino-benzidine and H_2O_2 in the substrate medium.

A careful application of the enzyme cytochemical technique has demonstrated the sites of many different enzymes in a wide variety of tissues. It is emphasized that a staining method suitable for localization of an enzyme in one tissue may not be suitable for other tissues. Since many factors may influence these studies, the results should be interpreted with caution. Artifacts may arise due to overincubation, and changes in the staining pattern may occur during dehydration and embedding.

ANALYSIS OF SUBCELLULAR FRACTIONS

Differential and density gradient centrifugations have been established procedures for studying cellular fractions in cell biology. An understanding of the physiology of the biologic system has been greatly facilitated by biochemical methods that can be applied to preparations containing highly dispersed cellular constituents. The tissue is homogenized, and the cells are effectively ruptured to release the particles. It is important that this initial process not cause a significant change to them. Systematic cell fractionation techniques can then be used to isolate specific, morphologically well defined cellular components. Although the purity of these fractions can be determined to some degree with the light microscope (LM), submicroscopic particles such as microsomal fractions cannot be examined in the LM. The mitochondrial fraction shows heavy contamination with fragments of other cytoplasmic components with this microscope.

It is difficult to correlate functional properties determined by biochemical studies with the specific morphologic structures at the subcellular level unless the fractions are homogenous and noncontaminated. The validity of biochemical results should be accompanied with careful morphologic studies with the TEM. Therefore, whenever particulate fractions in biologic tissues are investigated by biochemical methods, electron micrographs are required to support the biochemical results. Problems may be encountered when morphologic results are compared with those obtained by biochemical methods.

LOCALIZATION OF NUCLEOPROTEINS

Several methods using heavy metals such as iron, uranyl, and bismuth salts have been used to localize nucleoproteins. Although they enhance contrast and apparently have some specificity, they are usually somewhat unsuitable for OsO_4-fixed tissues.

A more successful method is to use aldehyde fixation instead of OsO_4. Reactive phosphate groups of the nucleic acids are then freed by acetylation of the amino groups in the protein. The tissue blocks are then treated with highly specific trivalent indium (indium trichloride in acetone). While other areas remain unstained in the thin section, a dense staining occurs with nucleic acid-containing structures.

In another method, small tissue blocks fixed in aldehydes are treated with specific nucleases or perchloric acid to dissolve nucleic acids. Tissues are then embedded in epoxy resins, and thin sections are stained with heavy metallic stains. Nucleic acid sites can then be identified by comparing with undigested, stained controls. Alternatively, thin sections can also be digested by this method. Thin sections of tissues embedded in water-soluble embedding media apparently give better results when digested by this technique.

Lipids of aldehyde-fixed frozen sections can also be extracted by ethanol or chloroform-ether. Posttreatment with OsO_4 or $KMnO_4$ is used to determine the binding site.

AUTORADIOGRAPHY

Electron autoradiography, a cytochemical method, consists of incorporating a radioactive material into a specimen and its detection by photographic techniques. The position and quantification of the radioactive material can then be done at the submicroscopic level. The TEM and LM autoradiography are based on the same principle. A thin section of the specimen containing the radioactive material is coated with a thin layer of photographic emulsion and stored in the dark for some weeks or months. Low energy electron emitting radioisotopes such as H^3 are commonly used. The radioactive material decays during the prolonged exposure time, and the ionizing radiation emitted from the source produce latent images in the emulsion. After photographic processing, the developed image is seen as grains of silver by electron microscopy (Fig. 16-2). The silver can grow in the form of irregular filaments, as clusters of silver specks, or as round particles around the latent image (Fig. 16-3).

The electron autoradiography has several advantages. The higher optical resolution of the TEM allows the visualization of fine structures and the silver grains. The higher electron autoradiographic resolution also provides a more accurate determination of the position of the radioactivity than is possible with light microscopy. More importantly, the great depth of focus of the TEM makes it possible to see the developed grain and the specimen in the same plane. Grain counting for quantitative work is also simplified.

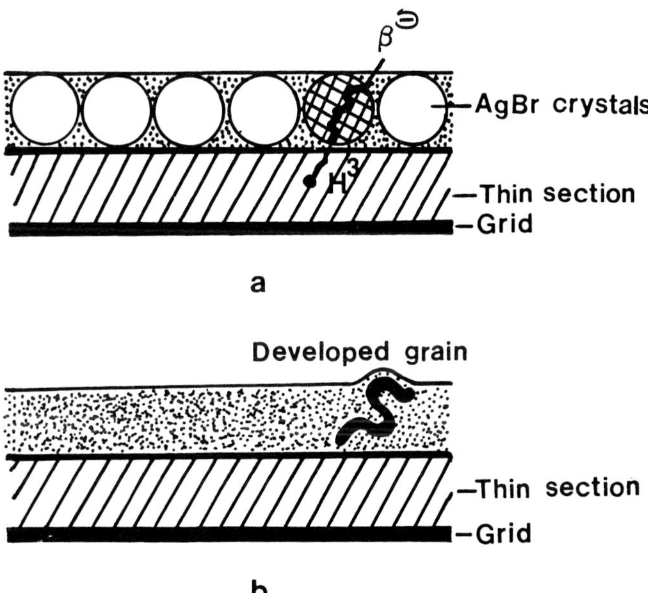

Figure 16-2. A schematic diagram of electron autoradiographic preparation. (a) During exposure. (b) After processing.

identification of cell components that carry antigenic determinants. The antigen-antibody reaction is observed by employing cytochemical reagents. The detector is usually incorporated into the antibody molecule for a covalent linkage.

The sensitivity of the methods is of paramount importance, but it should be realized that a decrease in specificity may occur when the sensitivity is increased. The nonspecificity in these reactions may occur due to nonspecific binding of the reagents, their byproducts, and contaminants. An immunologic nonspecific reaction may occur due to cross-reaction of specific antibodies with determinants other than the antigen. Impurities in the antigen used for immunization may also induce nonspecific antibody response. Appropriate controls should be included for monitoring the specificity of the results.

Ferritin-Conjugated Antibody Method

Originally developed by Singer (1959), this method is an extension of the immunofluorescence (fluorescent antibody) technique that has been most successfully used for localizing antigens in cells or tissues by light microscopy. For the TEM, a sufficiently electron dense material is used to covalently link the antibody molecule without inactivating it. Although ferritin, mercury, uranium, or iodine has been used for antibody labelling, the majority of studies have employed ferritin-labelled antibodies for this purpose.

Ferritin contains approximately 20% electron-dense iron in the form of micelles of ferric hydroxide-phosphate with a protein shell, the apoferritin. An electron-dense conjugate is prepared by labelling the specific IgG with ferritin (MW $\approx$ 650,000). Commercial ferritin is usually recrystallized several times before use. Originally, ferritin was first mixed in m-xylene diisocyanate or toluene 2,4-diisocyanate. It was then reacted with the specific IgG. This two-step procedure was used because the antibody is inactivated by diisocyanate.

Later, the bifunctional reagent p,p′-difluoro-m-m′-dinitrophenylsulfone (FNPS) was introduced for one-step conjugation of the antibody and ferritin. In this method, 5 mg of FNPS in 1 ml of cold acetone is added to a mixture containing 160 mg of IgG and 460 mg of ferritin dissolved in 2% cold aqueous sodium carbonate suspension to give a 4% protein solution. The mixture is continuously stirred for 24 hours at 4° C and then dialyzed extensively against normal saline solution. Approximately 20 to 25% antibody is usually labelled. The conjugate is centrifuged to remove a small precipitate. It can be purified by centrifugation, gel electrophoresis, or on a sephadex column. Purified ferritin-labelled antispecies-Igs are now commercially available.

The labelled antibody can be used to detect surface as well as intracellular antigens. Both direct and indirect (two-layer) methods similar to the immunofluorescence technique can be used. In the direct method, the antigen is stained directly with the labelled antibody. For the indirect method, the antigen is first reacted with the unlabelled specific antibody, e.g. Ig made in rabbits. Then an anti-rabbit Ig prepared in another species, e.g. antibodies against rabbit Ig prepared in goats, labelled with ferritin is added to react with the specifically attached rabbit Ig. The unlabelled antibody made in rabbits thus acts as an antibody to the homologous antigen on one side and then acts as an antigen to the labelled anti-Ig on the other. Although the direct method has been used for detecting the antigen in autoimmune diseases, e.g. lupus erythematosus, which is itself an Ig, the method is expensive, and the indirect method has been exclusively used.

For detecting intracellular antigens by the indirect method, small blocks of tissue are fixed in buffered formalin or glutaraldehyde (OsO_4 usually destroys antigenicity) for 15 to 30 minutes, washed and frozen. The frozen tissues can be cut (30 to 50 μm thick) on a cryostat if necessary. Cell cultures can be pelleted after fixation and subjected to one cycle of freezing and thawing. The frozen and thawed tissue or cell pellet is allowed to react with the unlabelled specific Ig for 30 minutes to 1 hour at room temperature or at 37° C. Fixation is necessary for making the cell permeable to antibody, and small breaks in the cell membrane caused by freezing and thawing facilitate the penetration of the antibody and the conjugate. Although the freezing process causes some damage to fine structures, most organelles are discernible.

After incubating with the unlabelled specific antibody, the tissue is washed, and ferritin-labelled anti-Ig is added. Following incubation and washing again, the tissue is postfixed in OsO_4, dehydrated, embedded, and thin sectioned. The stained sections are examined in the TEM for detecting the intracellular antigen. Electron-dense ferritin molecules are visible at the specific antigen-antibody reaction site (Fig. 16-4).

Enzyme-Conjugated Antibody Method

The overall contrast of ferritin-conjugated antibodies is very low, and some cell components labelled with this conjugate are often not detected. Furthermore, ferritin, due to its large MW, penetrates poorly, the antigen-antibody reaction is usually weak, and some labelled antibody may not reach the site of antigen.

The introduction of enzyme markers appears to overcome these difficulties. Although glucose oxidase or acid phosphatase has been used as a marker, horseradish peroxidase (PO) has been used most extensively for labelling. It has the advantage of having a small MW ($\approx$ 40,000), which allows its rapid penetration into the fixed tissue. After the enzyme attaches to the antigenic site via the specific

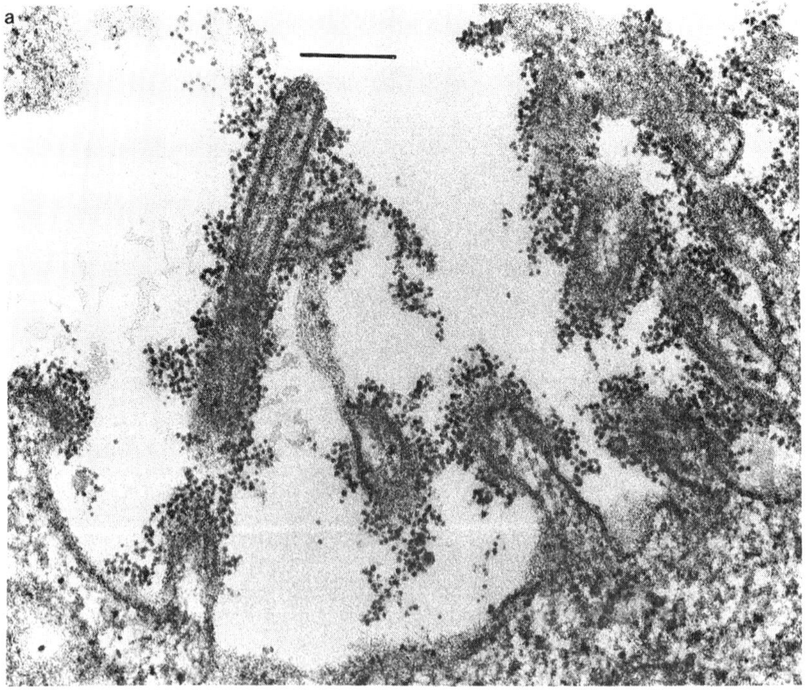

Figure 16-4. Ferritin-conjugated antibody technique. (a) Human respiratory syncytial virus grown in HeLa cells and processed by the indirect method. Bar = 100 nm. Courtesy of Dr. A. R. Kalica.

labelled antibody, a cytochemically detectable reaction for the enzyme is conducted with a suitable substrate. This yields a large number of visible product molecules per single molecule of antibody, and a very sensitive localization is achieved through the amplification of visible markers.

Peroxidase-antibody conjugate is prepared by mixing FNPS and IgG. To 50 mg of specific IgG, 50 mg of PO in 0.5 *M* cold carbonate buffer at pH 10 is added. Under gentle agitation, 0.25 ml of a 0.5% solution of FNPS in acetone is mixed. After continuous agitation for 6 hours at 4° C, the conjugate is dialyzed against phosphate buffered saline solution. It is centrifuged to remove a precipitate and then purified on a Biogel P300 column. The enzyme has an

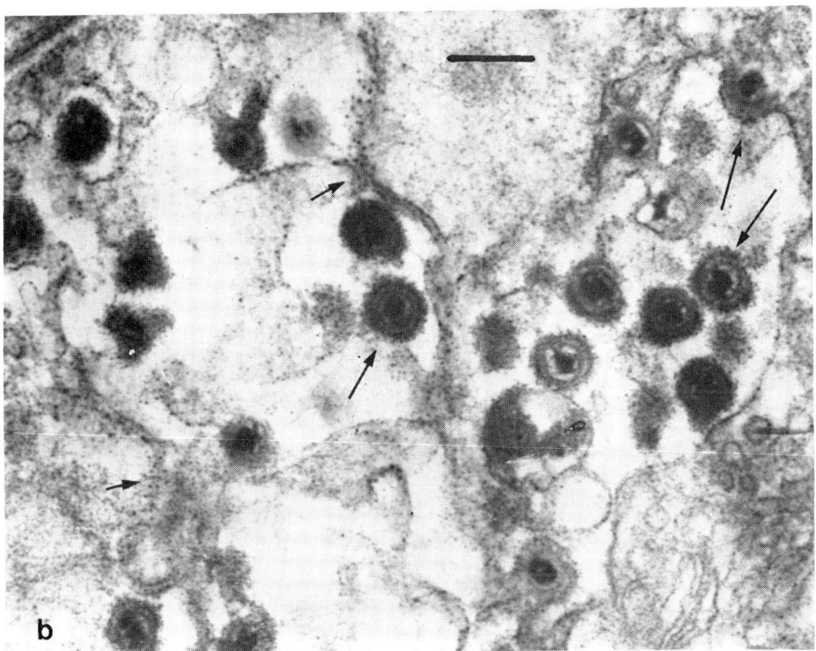

Figure 16-4. (b) Bovine embryonic kidney cells infected with a bovine herpesvirus strain DN-599 and processed by the indirect procedure. Note specific labelling of enveloped virions (long arrows) and cytoplasmic membrane (short arrows). Bar = 200 nm. From S. B. Mohanty, Immunoferritin and immune electron microscopic study of a bovine herpesvirus strain DN-599, *American Journal of Veterinary Research*, *36*:319, 1975. Courtesy of the American Veterinary Medical Association.

minute diffusion, the film is floated on water and picked up on grids.

The major disadvantage of the protein monolayer technique is that a protein is associated with the NA acid strands that is unsuitable for high-resolution work. The diameter of the DNA prepared by this method is ≈ 100Å compared to the normal DNA diameter of ≈ 20Å. Both double- and single-stranded NA molecules can be examined, but they are difficult to distinguish. Double-stranded DNA usually appears as finely curved filaments, whereas single-stranded DNA is heavily aggregated and curled. Formamide (75 to 95%) may be added to the protein-NA solution for observing the *denaturation loops* in denaturation mapping studies.

The Streaking Method

A carbon-coated grid is streaked along the surface of a DNA solution. Unbroken, straight, single NA molecules are picked up on the grid by this procedure. The direction of streaking is noted, and the excess fluid is blotted off. A DNA concentration of 1 to 3 μg/ml, 0.1 *M* ammonium acetate buffer, pH 6.1, streaking along the DNA solution for a distance of ≈ 2 cm at 2 mm/second are recommended.

The Adsorption Method

Carbon support films positively charged by treatment with quaternary ammonium salts are suitable for adsorbing NA molecules onto the grids. The technique has been modified to make the carbon-coated grids hydrophilic by glow discharge in a vacuum evaporator and treating them overnight with cationic detergent benzalkonium chloride (≈ 10^{-8} *M*). The grid is then touched onto the NA solution for picking up the macromolecules.

The Filter Method

A solution containing 10^{-3} to 5×10^{-4} μg total DNA or RNA is passed through a small Millipore® filter disk, and the strands are collected on the filter disk. A Teflon piston is recommended for passing the NA through the filter disk. The filter paper is then cut into small pieces, and the macromolecules may be stained at this time. An aluminum-coated grid is wetted with diethyleneglycol-dimethylether, and a piece of the filter paper is placed on the grid (which sticks to the grid) with the NA strands facing the film. The grid is then immersed in acetone for dissolving the filter paper completely.

The electron microscopy of NA strands has several important applications. The length of the DNA molecule can be measured by using a DNA of known length that is clearly distinguished from other DNA molecules present. In heteroduplex mapping studies, homology between two different DNA molecules or between DNA and RNA can be visualized. The gross features of nucleotide sequences of NA can be determined by denaturation mapping. The position of adenine-thymine-rich segments of DNA has been observed by this technique. Other defects such as nicks in DNA can also be seen. Denaturation can be accomplished by employing heat, high pH, or formamide. Techniques for observing DNA-protein complexes, e.g. DNA-RNA polymerase complex, have also been used.

SELECTIVE STAINING OF NUCLEOTIDES

Considerable progress has been made in selectively staining certain reactive groups within a molecule, especially the nucleotides in single-stranded DNA molecules. These marked groups have then been identified by high-resolution electron microscopy. Problems encountered in this technique include maintenance of the structure of the molecule, preparation of suitable low-noise support films, deposition of the molecules on the grid, selectivity of the staining reagent, and identification of selectively stained sites.

It is critical that the support film used for these studies have minimal background structure so as not to obscure the marked sites. Single crystal graphite film and very thin, amorphous aluminum oxide films have been promising in this respect. The unbroken, untangled single NA molecules are best deposited by the adsorption method, in which the surface of the support film is positively charged, or by the filter method (*see* in this Chap.).

For intramolecular staining, a few reagents have been found to stain the nucleotides selectively. Diazotized 2-amino-p-benzene disulfonic acid at pH 9 appears to be specific for guanine. Cytosine-specific reactions have been obtained by staining with semicar-bazides at pH 4.2 or with acyl hydrazides, e.g. 3,4-dicarboxybenzoyl hydrazine. Thymine has been selectively stained with OsO_4, and K_2PtCl_4 appears to react primarily with adenine residues. For selective staining of proteins, PTA has been used to demonstrate guanidine bonds in collagen.

With refined staining methods and/or high-resolution scanning transmission electron microscopy (*see* Chap. 17), it is believed that the sequence of nucleotides in NAs will be more clearly recognized in the future.

HIGH VOLTAGE ELECTRON MICROSCOPY

The term *high voltage electron microscope* (HVEM) is commonly used for a conventional transmission electron microscope (TEM) that operates at an accelerating voltage of 200 KV or higher. There are approximately 42 HVEMs having accelerating potentials $\geqslant$ 500 KV throughout the world; 12 of them are in the U.S.A. Some of these microscopes operate at 1,000 KV (1 MV), but a few are designed for 3 to 5 MV operation. Although HVEMs have been mainly used for research in material science, metallurgy in particular, biologic applications of these instruments are on the rise. Owing to the unusually high cost of these sophisticated instruments, they are mostly owned by government agencies or industries and are frequently operated on a time-sharing basis. Sponsored by the National Institutes of Health, one 1 MV HVEM for biologic research has been installed at the University of Colorado, Boulder, and one at the University of Wisconsin, Madison. Another similar instrument is used at the New York State Department of Health, Albany. Several 3 MV HVEMs are in operation throughout the world including the laboratory of DuPouy, Tolouse, France.

The rationale for using the HVEM is to improve the overall resolving power of the instrument, which is expressed by the following equation:

$$R = K\,(C_s\,\lambda^3)^{1/4}$$

where R = overall resolving power of the instrument; K = constant; C_s = spherical aberration constant of the objective lens; and λ = wavelength of the incident electron. Theoretically, therefore, an HVEM should provide a significantly higher resolution due to a very small wavelength (*see* de Broglie's postulate, Chap. 1). The wavelength of an electron at 100 KV is 0.038 Å, but it is 0.0012 Å at 1 MV and 0.0007 Å at 3 MV. A theoretical resolution of 1 Å (atomic resolution) is possible with a 1 MV or 3 MV instrument.

The principal advantages of an HVEM are (a) increased penetration that allows examination of thicker specimens, (b) improved resolution mainly due to decreased chromatic aberration, and (c) reduced ionization damage. The disadvantages are (a) reduced contrast, and (b) damage due to displacement of atoms, and (c) reduced efficiency of the viewing screen and photographic emulsion.

Increased Specimen Penetration

This is probably the most significant advantage of the HVEM over a conventional 100 KV TEM. Operating at a higher KV makes thick specimens more transparent to the beam than at lower KV. The maximum specimen thickness that can be penetrated by a 100 KV beam is $\approx$ 0.1 μm (1,000 Å); at 500 KV the penetration is 2.4 $\times$ more, and $\approx$ 3 to 10 $\times$ higher penetration can be obtained at 1 to 3 MV. This greatly facilitates the three-dimensional viewing of thick specimens. At present, however, it appears that the optimal thickness that can be profitably examined in a 1 to 3 MV HVEM is 3 μm. A serious problem of superimposition of internal structures occurs (due to large depth of field and focus), and stereoscopy is required for most thick specimen viewing. The examination of very thick specimens is limited by temperature rise, drift, and radiation damage rather than due to inadequate transmission or poor resolution caused by chromatic aberration.

Increased Resolution

In addition to a reduction in wavelength of high-velocity electrons, which improves the resolving power of HVEMs, the resolution of thick specimens improves markedly due to decreased chromatic aberration. Compared to electrons accelerated at lower KVs, the scattering of high-energy electrons in an HVEM causes less energy loss (*see* Chap. 4). This effect is probably the single most important factor in improving the resolution of thick specimens in HVEMs. A higher resolution can also be expected due to smaller spherical aberration at a higher accelerating voltage.

Beam Damage

Ionization and displacement of atoms caused by the electron beam are mainly responsible for specimen damage. The ionization damage is caused by displacement of electrons from the specimen. Although it does not produce structural changes in metallic specimens, it frequently causes permanent changes in covalently bonded biologic structures. A *critical exposure* of the specimen to the beam is required before a certain structural change occurs, and this exposure is dependent on the type of specimen and incident electron energy. The electrons in an HVEM travel with an extremely high velocity; for 1 and 3 MV electrons, the velocity is ≈ 0.94 and 0.99 times the velocity of light, respectively. Thus, the electron-specimen interaction time is reduced, and the transfer of energy of the electron is smaller. The resultant energy loss of the electron is also reduced. This causes less ionization damage because the probability of single electron excitation (ionization damage) is inversely proportional to the square of the incident electron velocity.

Sublimation of staining materials, breaks, and thermal drift occur in thick sections over 0.1 μm thick at 75 or 100 KV. These phenomena are not noticed with HVEMs at 500 to 800 KV even when the beam is brought to crossover. This indicates that less energy is transferred to the specimen by the high-energy beam, and less ionization damage occurs.

The problem of radiation damage due to displacement of atoms, however, may be serious in biologic specimens in an HVEM. The incident high-energy electron can transfer sufficient energy to an atom of the specimen and displace it from its original position. Although displacement of atoms has been demonstrated in metals, data from biologic specimens are lacking. The probability of ionization damage is far greater than the damage caused by displacement of atoms, and it may well be that the latter damage is negligible in biologic specimens in an HVEM.

Theoretically, for a specimen of given thickness, the beam heating effect should be 3 times less at 1 MV than at 100 KV, but this effect has been a problem for very thick biologic specimens (thicker than 3 μm) in HVEM (*see* in this Chap.).

Instrumentation

The HVEM column is essentially a scaled-up version of the

conventional TEM column (Fig. 16-8). High voltage electron beams require larger and more powerful magnetic lenses to focus them. A typical microscope column consisting of 3 or 4 imaging lenses, the viewing and recording chambers is ≈ 3 m high. Each lens of most 1 to 3 MV HVEMs is ≈ 80 cm in dia, 50 cm high, and weighs ≈ ½ ton. The overall height of most of these instruments exceeds 10 m. Almost half of it is occupied by the high voltage generator and stabilizer, and these are mounted on a subfloor above the column. The high voltage is generated and stabilized in a linear accelerator tank containing an insulating gas under pressure (in a few HVEMs, this is air insulated). All HVEMs require heavy shielding for protection against x-rays. Currently, AEI, Hitachi, and JEOL manufacture extremely expensive and sophisticated HVEMs.

Contrast, Astigmatism, and Image Recording

Phase contrast does not occur in specimens ranging in thickness

Figure 16-8. The Hitachi 3 million volt electron microscope. Courtesy of Hitachi Ltd.

from 0.2 μm and higher. Therefore, the amplitude contrast, which decreases with increasing accelerating voltages, is usually low in HVEMs. A thin plastic plate for a fluorescent screen has been used in some HVEMs in place of a metallic base plate. This apparently reduces the back scatter of electrons from the screen base and improves contrast. Astigmatism correction in HVEMs is usually a lengthy procedure and is done by trial and error methods. The response of the fluorescent screen and the photographic emulsion usually decreases to high-energy electrons in HVEMs. A thicker emulsion is usually used, and a fast film, Kodirex, has been useful for high resolution photography. A *through-focal series* is usually taken for recording the image.

Biologic Applications

Although HVEMs operated at 1 to 3 MV can provide lattice resolution of 2 to 3 Å, the most significant benefit derived from HVEMs is the increased specimen penetration. This is illustrated in Figures 16-9 to 16-11. Superimposition of structures in thick specimens poses a special problem, and stereomicroscopy is generally used for their three-dimensional viewing. Thick specimens are usually block stained, and selective staining can be used to enhance contrast of desired structures.

There have been some doubts that living specimens can be examined in HVEMs. However, bacteria and spores have been claimed to have survived the examination in HVEMs. The use of HVEMs could be potentially extended to electron autoradiography with thick sections, and high voltage scanning transmission electron microscopy (STEM) is possible. The value of HVEMs in dark-field electron microscopy for examining unstained thin specimens has been demonstrated, and examination of *wet cells* in environmental chambers with HVEMs has been encouraging (*see* in this Chap.).

Extensive research on high voltage microscopy with 1 to 5 MV HVEMs is being conducted in England, France, Japan, and the U.S.A. In spite of the cost disadvantages, the use of these microscopes in biology, although still in its infancy, appears to be steadily increasing.

DARK-FIELD ELECTRON MICROSCOPY

A dark-field image is formed when unscattered electrons are

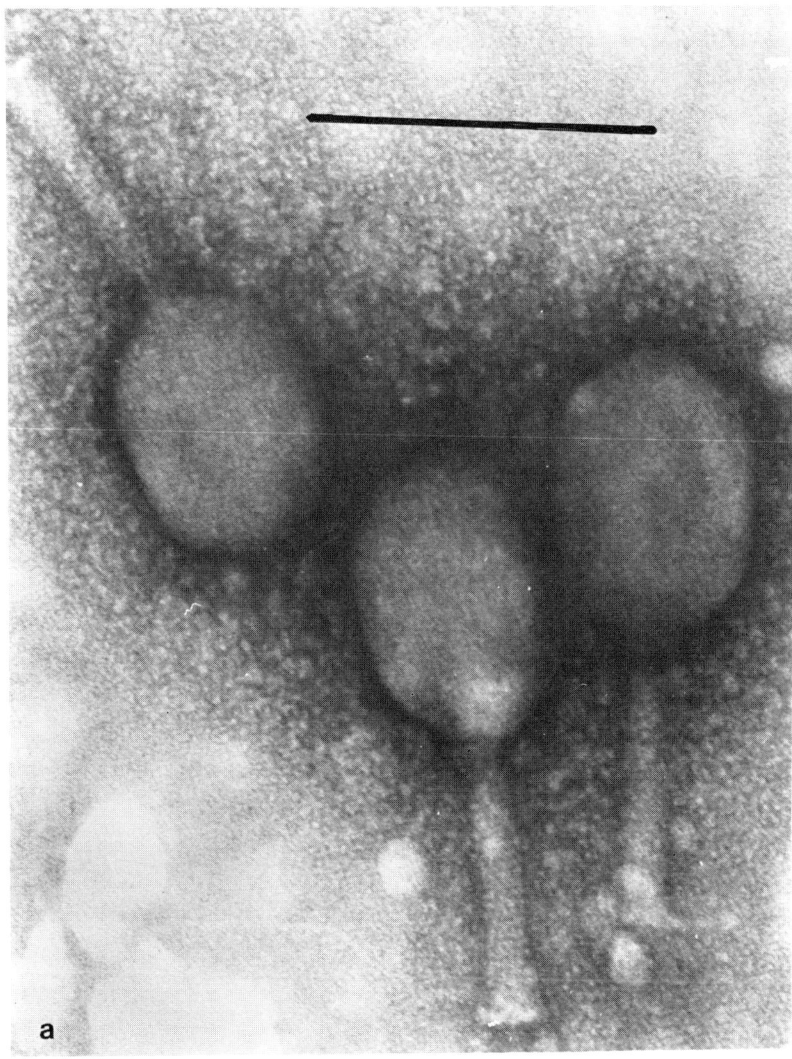

Figure 16-9. T_4 bacteriophage negatively stained. (a) Examined at 100 KV.

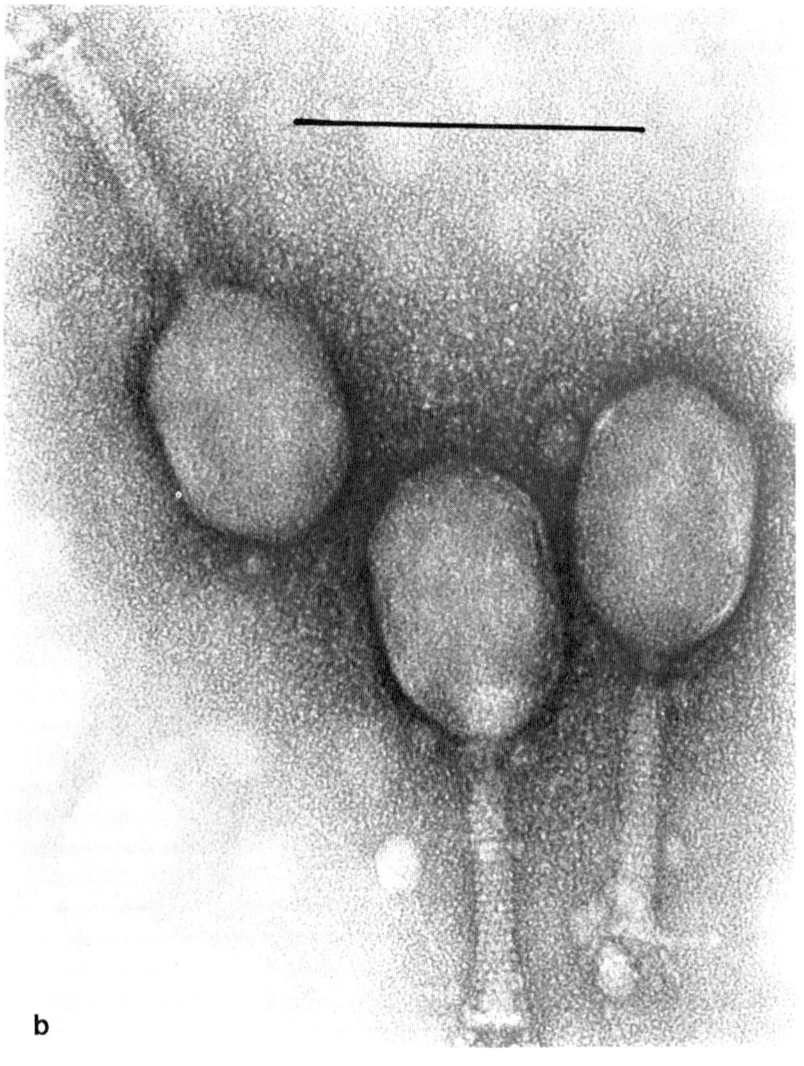

Figure 16-9. (b) Examined at 200 KV. Bar = 100 nm. Courtesy of Hitachi Ltd.

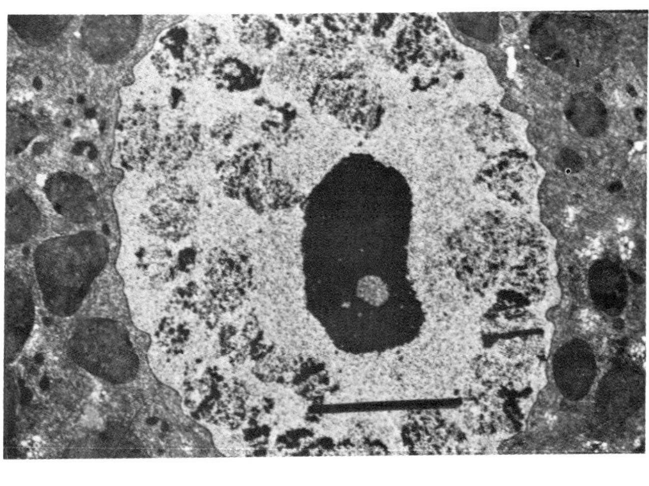

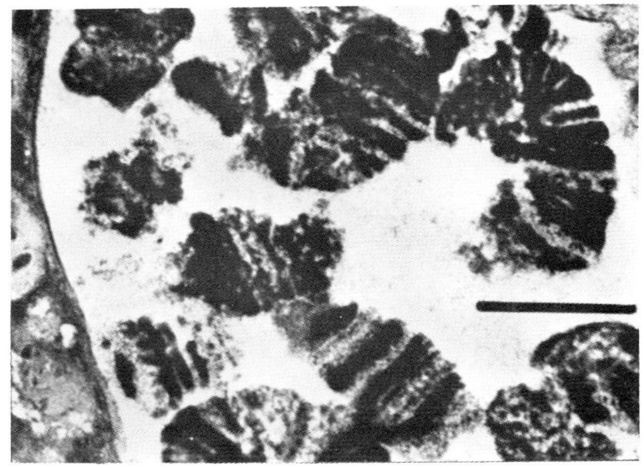

Figure 16-10. A nucleus of a drosophila salivary gland. (a) An 80 nm-thick section examined at 75 KV. Giant chromosomes cannot be seen in this thin section. Bar = 2 μm. (b) Giant chromosomes and band patterns are observed at 800 KV with a thick section (1 μm thick) cut from the same block as in Figure 16-10(a). Bar = 2 μm. From K. Hama, High voltage electron microscopy. In J. K. Koehler (Ed.), *Advanced Techniques in Biological Electron Microscopy*, 1973. Courtesy of Springer-Verlag, New York.

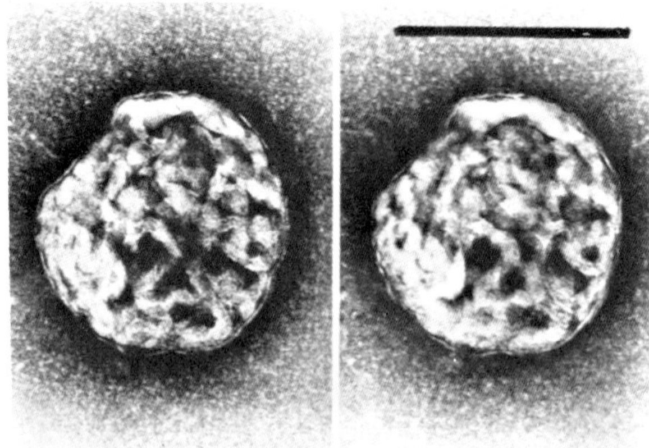

Figure 16-11. A stereo pair of negatively stained mitochondrion showing tubular cristae. Examined at 800 KV. Bar = 1 μm. From K. Hama, High Voltage electron microscopy. In J. K. Koehler (Ed.), *Advanced Techniques in Biological Electron Microscopy*, 1973. Courtesy of Springer-Verlag, New York.

intercepted by an opaque mask so that electrons are scattered only by the specimen. The main beam passing through the specimen is usually deflected or intercepted, and it does not reach the screen. The electrons scattered only by the specimen pass through the imaging lens system to form an image of reverse contrast (opposite to bright field imaging). In most cases, the image formed by the dark field usually contains more information than that formed by the bright field.

Methods of Obtaining Dark-Field Images

There are 4 methods for obtaining dark-field images. These are shown in Figure 16-12 and are described below.

Displacement of Objective Aperture (Aperture Shift)

This is the easiest method; electron microscopists commonly see dark-field images due to displacement of objective aperture. In this method, the objective aperture is moved to one side so that it intercepts the main beam but allows the electrons scattered by the specimen to enter the imaging system asymmetrically. The spherical

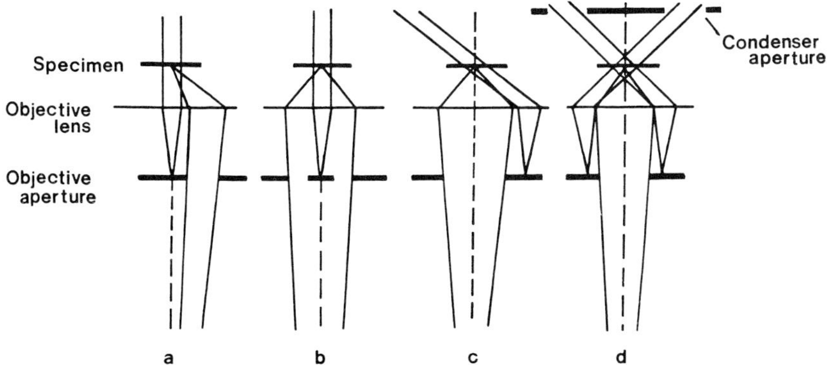

Figure 16-12. Methods of dark-field image formation. (a) Displaced objective aperture. (b) Beam-stop aperture (strioscopy). (c) Beam-tilt. (d) Conical illumination with condenser aperture. Adapted from J. Dubochet, High-resolution dark-field electron microscopy. In M. A. Hayat (Ed.), *Principles and Techniques of Electron Microscopy, Biological Applications,* Vol. 3. New York, Van Nostrand, 1973.

aberration is increased with this method, and a good quality image cannot be obtained by this technique.

Beam-Stop (Strioscopy, Axial Illumination)

A special objective aperture into which a thin wire is welded acts as the central stop for the main beam. The central part of the thin wire can also be melted to produce a drop that can be used as a screen at the center. The contrast is greatly improved by this method, particularly in HVEMs.

Beam-Tilt

For this method, the illuminating beam is tilted so that it does not pass along the axis of the objective lens. Although the intensity on the screen is usually less than the first two methods described, the scattered beam enters the imaging system fairly symmetrically, chromatic aberration is reduced to the minimum, and a high resolution is obtained. Ottensmeyer has obtained excellent results with unstained biomacromolecules by this method (Fig. 16-13). Beam-tilt devices are equipped in several microscopes, and this type

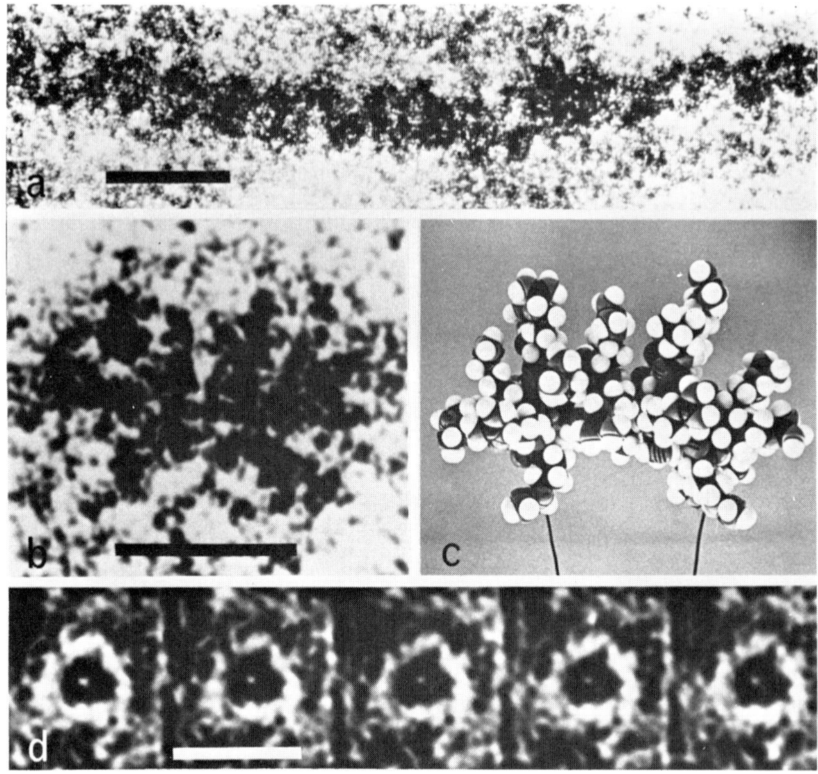

Figure 16-13. High-resolution dark-field electron micrographs. (a) γDNA. (b) Protamine. (c) Protamine model. (d) Valinomycin. Bar = 2 nm. Courtesy of Dr. F. P. Ottensmeyer.

of dark field can be easily achieved with them. A double-deflector system is usually incorporated into the second condenser lens for this purpose.

Conical Illumination by Annular Condenser Aperture

This method of obtaining dark field has several advantages. It provides good illumination and reduces asymmetry of the beam. A condenser annular aperture of $\approx$ 1 mm in diameter is used for this purpose, and the aperture can be obtained from several electron microscope manufacturers.

For obtaining good dark-field images, the TEM is first aligned in bright field with a thin specimen but without the objective aperture; dark field is then obtained by one of several methods. Since the main beam is intercepted in dark-field electron microscopy, it is critical that the beam intensity be high. The lack of intensity and difficulty in focusing are probably the most important problems encountered in dark-field operation. Astigmatism correction is also difficult, but it can be avoided by first correcting the residual astigmatism in bright field and by using clean apertures. The specimen must be thin, and it can be stained or shadowed, but unstained specimens have given good contrast and high resolution. Carbon is most commonly used as the support film, and it should not be thicker than 30 to 50Å.

Excellent high-resolution micrographs have been obtained by dark-field electron microscopy. These show the double helical nature of DNA, cloverleaf shape of t-RNA, globular structure of ribonuclease with a small cleft (all predicted by x-ray diffraction studies), and visualization of single heavy atoms. The potential of dark-field electron microscopy appears to be greatest for high-resolution work, especially with HVEMs.

ENVIRONMENTAL SPECIMEN CHAMBERS

Most biologic specimens are examined in a dehydrated state in the TEM that operates with a minimum vacuum of 10^{-4} torr. This may cause structural changes in these specimens, and indeed, x-ray and electron diffraction studies with periodic biologic structures indicate that their diffraction patterns disappear as soon as they are dehydrated. Attempts have been made for transmission electron microscopy of specimens by surrounding them in a liquid or highly humid gaseous atmosphere at atmospheric pressure. Such devices could allow the observation of inner structures of living specimens, dynamic movement of cellular organelles, and biochemical activities.

A liquid must be surrounded by its saturated vapor to maintain it in equilibrium in the TEM. The vapor pressure is dependent on the properties of the liquid and the temperature of observation. Saturated vapor pressure, however, is considerably higher than the vacuum pressure in the TEM. This requires the construction of pressure chambers that would not burst open when introduced into the vacuum. The chamber should also allow high electron transmittance. Special specimen holders that sustain a nonvacuum environment around the specimen are variously called *environmental devices, hydration chambers, wet cell chambers,* or *environmental specimen chambers.*

Types of Hydration Chambers

Two types of environmental devices are currently used in conventional TEMs as well as in HVEMs.

Window- or Membrane-Type Chambers

This type of device consists of electron-transparent membranes above and below the specimen that can withstand a static pressure difference. This design encapsulates and completely isolates the specimen and its environment from the vacuum in the microscope, but a leak may develop in this sealed system.

Conventional objective aperture disks serve as the best window supports, and apertures of 50 to 100 μm diameter are commonly used. The air slab is usually 5 to 20 μm thick bounded at opposite faces by a three-layer film $\approx$ 600 Å thick (Fig. 16-14a). The apertured disk is first coated with collodion or formvar (150 to 300 Å), and a thin layer of carbon (50 to 200 Å) is evaporated onto it for providing improved thermal and electrical conductivity. A thin layer of silicon monoxide ($\approx$ 100 Å) is then evaporated onto the plastic surface exposed to the high pressure region. This increases the life expectancy of the chamber and prevents O_2 attack on the carbon. The incident electron beam travels through the two windows, the specimen, some water, and humid air. This type of cell is suitable for containing liquids and gases, but the window material scatters electrons. A few window chambers are open to the exterior through a fine tubing so that water can be externally introduced into the cell.

Differentially Pumped or Throttling Aperture Chambers

The high-pressure region separated from main vacuum in this system is maintained by using 2 sets of apertures, $\approx$ 50 μm diameter, above and below the specimen (Fig. 16-14b). A space of 1 to 2 mm is provided between each set of apertures. The resulting small gas leak

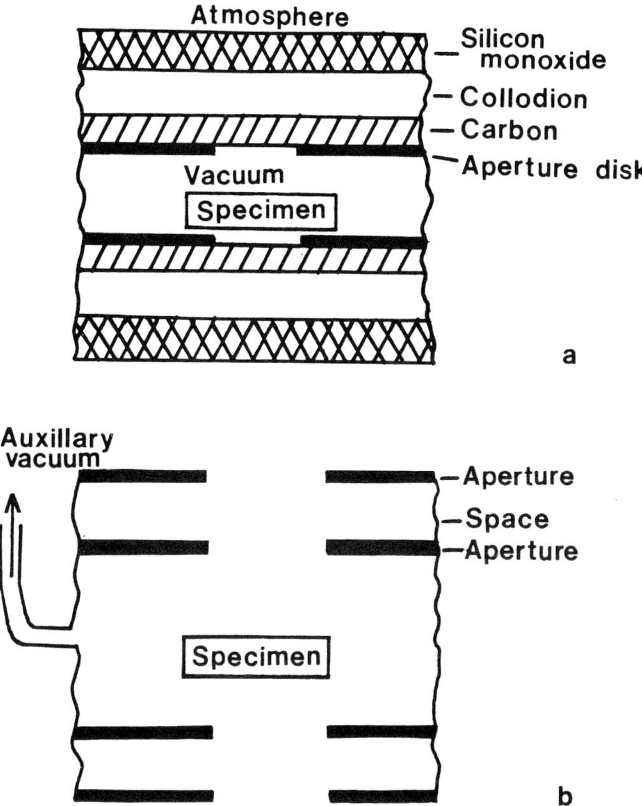

Figure 16-14. Schematic diagrams of environmental chambers. (a) Window type. (b) Throttling aperture type (differentially pumped).

through the apertures within the cell is then pumped by an auxiliary pumping system so that the vacuum in the main column is not degraded. The specimen is usually in contact with a 24-torr pressure of water vapor. These devices are simpler to construct than the window type, and entrance and exit apertures are easily aligned. A pressure gradient across the apertures is maintained by having a source of water vapor that feeds the specimen and then by differentially pumping the lower pressure side of the aperture. A resolution of 40Å has been reported with some test specimens with this type of device.

Resolution, Contrast, and Radiation Damage

The resolution and contrast are always worse with hydration chambers than when similar specimens are examined in high vacuum. The resolution in these devices is mainly limited by chromatic aberration. This is especially worse with window chambers where an additional scattering film is placed in the path of electrons emerging from the specimen. The contrast of wet, unstained, and unfixed specimens is also poor. A multiple scattering occurs in differentially pumped chambers, which gives rise to loss of contrast. Although various cytologic structures have been observed in wet, unstained cells, the resolution and contrast of these specimens have so far been disappointing. The specimens examined in environmental devices are usually thick, and it is doubtful that the resolution of wet specimens would approach that obtained by thin sectioning.

By the very nature of the technique, the electron and x-ray radiation damage to the specimen is always moderate to severe. These damages could be minimized by using a helium gaseous environment saturated with water vapor and by maintaining the chamber at as low a pressure as possible. Improved recording systems, e.g. image intensifiers, with HVEMs could also be used to reduce radiation-induced structural damage.

Maintenance of Humid Conditions

It is difficult to maintain the specimen in a hydrated state in the chamber. The resolution and contrast deteriorate rapidly when the specimen is bathed in water films. To keep the specimens wet, biologists generally pass light atomic weight gas, e.g. helium, saturated with water vapor around the specimen. This reduces the rate of specimen dehydration to a very low level. A very high relative humidity is required, and the diffraction pattern of catalase crystal plates is obtained only when the specimen is kept at a relative humidity of greater than 93%.

A variety of biologic specimens including bacteria, spores, cultured cells, catalase crystals, and platelets have been examined by this technique (Fig. 16-15). Some workers have used gold grids to overcome the toxicity of copper grids.

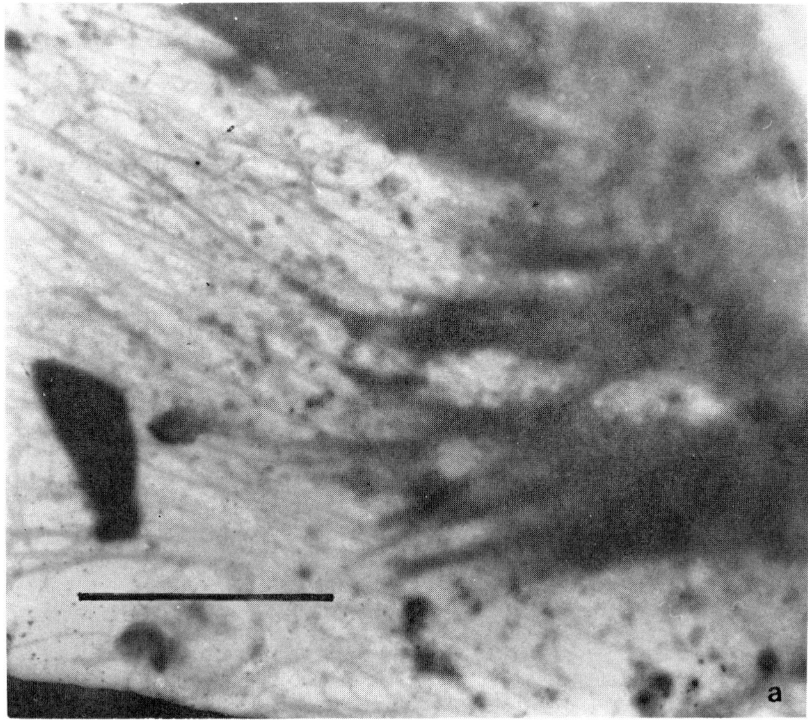

Figure 16-15. (a) Cultured mouse fibroblasts examined in a hydration chamber with a 1 million volt electron microscope. Bar = 5 μm. From D. F. Parsons, I. Uydess, and V. R. Matricardi, High voltage electron microscopy of wet whole cells. Effect of different cell preparation methods on visibility of structure. *Journal of Microscopy, 100*:143, 1974. Courtesy of the Royal Microscopical Society.

Although several *E. coli* cells have been seen to divide in the hydration chamber during electron microscopic examination, and other bacteria have been claimed to have survived after this examination, these reports have been highly controversial. Some investigators believe that a high resolution cannot be expected in a wet atmosphere. Even then, biologists still hold out the tantalizing hope that living cells can some day be examined by this technique. The value of incorporating these devices in HVEMs has been demonstrated. Clearly, the increased penetration with HVEMs (1 to 3 MV) holds promise for this technique, and the use of HVEMs

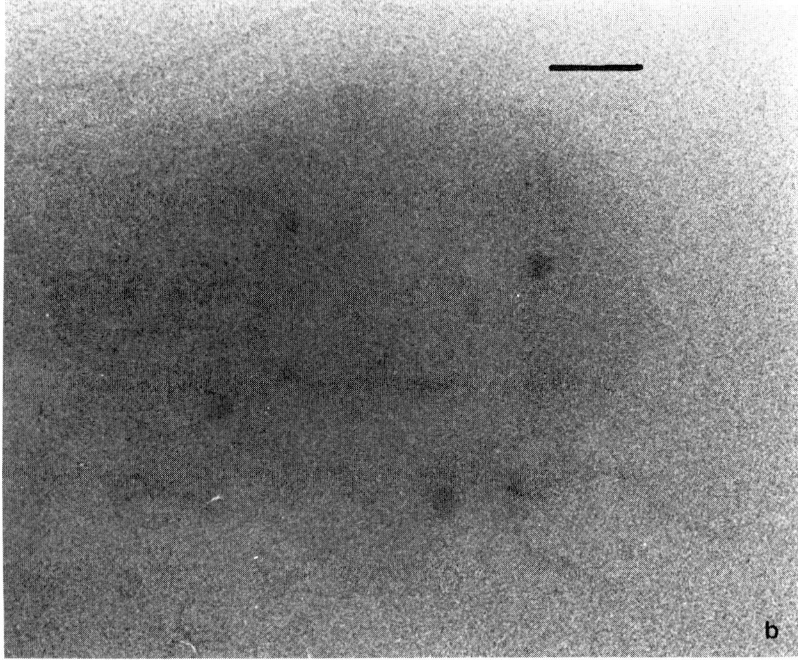

Figure 16-15. (b) Thrombin-treated human platelets with dense bodies at 5° C. Examined in a hydrated state at 100 KV. From J. L. Costa and S. W. Hui, *In situ* observation of dense-body release from hydrated human platelets. *Biophysical Journal, 19:*307, 1977. Courtesy of the Rockefeller University Press, New York.

would dominate this field. Some workers also predict that environmental chambers may be incorporated into the most sophisticated STEMs in the future.

SELECTED BIBLIOGRAPHY

General

Glauert, A.M.: Section staining, cytology, autoradiography and immunocytochemistry for biological specimens. In Kay, D.J. (Ed.): *Techniques for Electron Microscopy*, 2nd ed. Philadelphia, F.A. Davis, 1965.

Meek, G.A.: *Practical Electron Microscopy for Biologists*, 2nd ed. London, Wiley, 1976.

Wischnitzer, S.: *Introduction to Electron Microscopy*, 3rd ed. New York, Pergamon, 1981.

Specific Articles

Localization of Enzymes (Enzyme Cytochemistry)

Hayat, M.A. (Ed.): *Electron Microscopy of Enzymes: Principles and Methods*, vols. 1 and 2. New York, Van Nostrand, 1973 and 1974.

Holt, S.J. and Hicks, M.: Combination of cytochemical staining methods for enzyme localization with electron microscopy. In Harris, R.J.C. (Ed.): The Interpretation of Ultrastructure, vol. 1, New York, Academic Press, 1962.

Mori, M. and Novikoff, A.B.: Induction of pinocytosis in rat hepatocytes by partial hepatectomy. *J Cell Biol*, *72*:695, 1977.

Novikoff, A.B.: Some contributions of phosphatase and 3,3′-diamino-benzidine cytochemistry to cell biology. *Proc EMSA: 35th Annual Meeting*, 420, 1977.

Analysis of Subcellular Fractions

Deter, R.L.: Electron microscopic evaluation of subcellular fractions obtained by ultracentrifugation. In Hayat, M.A. (Ed.): *Principles and Techniques of Electron Microscopy: Biological Applications*, vol. 3. New York, Van Nostrand, 1973.

Localization of Nucleoproteins

Gautier, A.: Ultrastructural localization of DNA in ultrathin sections. *Int Rev Cytol*, *44*:113, 1976.

Swift, H.: Nucleoprotein localization in electron micrographs: Metal binding and radioautography. In Harris, R.J.C. (Ed.): *The Interpretation of Ultrastructure*, vol. 1. New York, Academic Press, 1962.

Autoradiography

Salpeter, M.M. and Bachmann, L.: Autoradiography. In Hayat, M.A. (Ed.): *Principles and Techniques of Electron Microscopy: Biological Applications*, vol. 2. New York, Van Nostrand, 1972.

Salpeter, M.M. and McHenry, F.A.: Electron microscope autoradiography. In Koehler, J.K. (Ed.): *Advanced Techniques in Electron Microscopy*. New York, Springer-Verlag, 1973.

Yamamoto, T. and Shahrabadi, M.S.: Enzyme cytochemistry and autoradiography of adenovirus-infected cells as determined with the electron microscope. *Can J Microbiol*, *17*:249, 1971.

Immunologic Methods

Anderson, T.F.: Negative staining and its use in the study of viruses and their serologic reactions. In Harris, R.J.C. (Ed.): *The Interpretation of Ultrastructure*, vol. I. New York, Academic Press, 1962.

Kapikian, A.Z., Feinstone, S.M., Purcell, R.H., Wyatt, R.G., Thornhill, T.S., Kalica, A.R., and Chanock, R.M.: Detection and identification by immune electron microscopy of fastidious agents associated with respiratory illness, acute non-bacterial enteritis, and hepatitis A. *Perspect Virol*, *9*:9, 1975.

Mohanty, S.B.: Immunoferritin and immune electron microscopic study of bovine herpesvirus strain DN-599. *Am J Vet Res*, *36*:319, 1975.

Singer, S.J.: Preparation of an electron dense antibody conjugate. *Nature*, *183*:1523, 1959.

Sternberger, L.A.: Enzyme immunocytochemistry. In Hayat, M.A. (Ed.): *Electron Microscopy of Enzymes: Principles and Methods*, vol. 1, New York, Van Nostrand, 1973.

Electron Microscopy of Nucleic Acid Molecules and Selective Staining of Nucleotides

Kleinschmidt, A.K., and Zahn, R.P.: Über Desoxyribonucleinsäure-Molekeln in Protein-Mischfilmen. *Z Naturforsch 14b*:770, 1959.

Koller, T., Beer, M., Müller, M., Mühlethaler, K.: Electron microscopy of selectively stained molecules. In Hayat, M.A. (Ed.): *Principles and Techniques of Electron Microscopy: Biological Applications*, vol. 3, New York, Van Nostrand, 1973.

Moore, C.L.: The electron microscopy of ribonucleic acids. In Griffith, J.D. (Ed.): *Electron Microscopy in Biology*, vol. 1. New York, Wiley, 1981.

Thomas, J.O.: Electron Microscopy of DNA. In Hayat, M.A. (Ed.): *Principles and Techniques of Electron Microscopy: Biological Applications*, vol. 9. New York, Van Nostrand, 1978.

High Voltage Electron Microscopy

DuPouy, G.: Electron microscopy at very high voltages. *Adv Opt Electron Microsc*, *2*:167, 1968.

Fotino, M.: Experimental studies on resolution in high voltage transmission electron microscopy. In Griffith, J.D. (Ed.): *Electron Microscopy in Biology*, vol. 1. New York, Wiley, 1981.

Hama, K.: High voltage electron microscopy. In Koehler, J.A. (Ed.): *Advanced Techniques in Biological Electron Microscopy*. New York, Springer-Verlag, 1973.

Humphreys, C.: High voltage electron microscopy. In Hayat, M.A. (Ed.): *Principles and Techniques of Electron Microscopy: Biological Applications*, vol. 6, New York, Van Nostrand, 1976.

Dark-Field Electron Microscopy

Andrews, D.W. and Ottensmeyer, F.P.: Imaging a protein α helix by dark-field electron microscopy. *Proc EMSA, 39th Annual Meeting*, 34, 1981.

Dubochet, J.: High resolution dark-field electron microscopy. In Hayat, M.A. (Ed.): *Principles and Techniques of Electron Microscopy: Biological Applications*, vol. 3. New York, Van Nostrand, 1973.

Hashimoto, H., Kumao, A., Hino, K., Endoh, H., Yostumoto, H., and Ono, A.: Visualization of single atoms in molecules and crystals by dark-field electron microscopy. *J Electron Microscopy*, *22*:123, 1973.

Ottensmeyer, F.P.: Macromolecular fine structure by dark-field electron microscopy. *Biophys J*, *9*:1144, 1969.

Ottensmeyer, F.P., Whiting, R.F., and Korn, A.P.: The structure of macromolecules at 5 angstroms. *Proc EMSA, 33rd Annual Meeting*. 14, 1975.

Environmental Specimen Chambers

Allinson, D.L.: Environmental devices in electron microscopy. In Hayat, M.A. (Ed.): *Principles and Techniques of Electron Microscopy: Biological Applications*, vol. 3, New York, Van Nostrand, 1975.

Costa, J.L. and Hui, S.W.: *In situ* observation of dense-body release from hydrated human platelets. *Biophys J*, *19*:307, 1977.

Parsons, D.F., Uydess, I., and Matricardi, V.R.: High voltage electron microscopy of wet whole cells. Effect of different wet cell preparation methods on visibility of structures. *J Microscopy*, *100*:153, 1973.

Chapter 17

SCANNING, SCANNING TRANSMISSION ELECTRON MICROSCOPY, AND X-RAY MICROANALYSIS

So far, we have discussed the conventional transmission electron microscope (TEM), the extensive biologic use of which was started in the 1950s. There are other instruments related to the TEM; we will discuss them and consider their biologic applications in this chapter.

During electron microscopy, the incident electron beam interacts with the specimen in various ways, and the most important of these interactions are shown diagramatically in Figure 17-1. Most of the electron beam specimen interactions have been discussed in Chapter 4. Auger electrons (named after the discoverer) are low-energy electrons emitted very near the surface for filling up the holes left by secondary electron emission to dissipate excess energy in the stabilization process. Cathodoluminescence is the emission of photons (UV, visible light, or infrared) in some materials for dissipating this energy. Signals generated by the electron-specimen interaction can be collected for image formation by different types of EMs.

SCANNING ELECTRON MICROSCOPY

The scanning electron microscope (SEM) is particularly useful for examining the surfaces of specimens, and it provides a three-dimensional image of the surface topography. Along with an x-ray detector, it can give useful information about the chemical composition of the specimen and other characteristics of biologic objects.

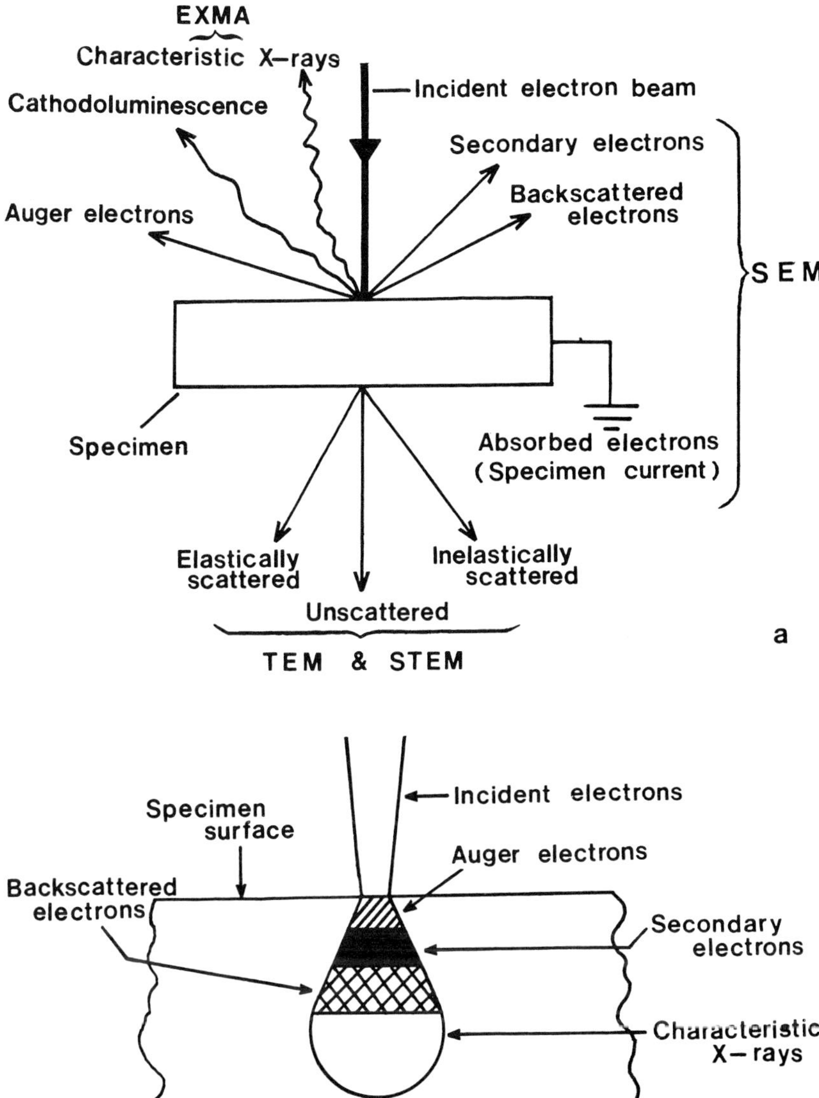

Figure 17-1. Electron-specimen interaction. (a) Most important signals generated by this interaction that can be collected for various types of electron microscope image formation. SEM = scanning electron microscope; EXMA = electron X-ray microanalyzer; TEM = transmission electron microscope; STEM = scanning transmission electron microscope. (b) Cross section of the electron excitation volume showing zones from which signals may be collected.

The useful resolution of the SEM, although not comparable to the TEM, has improved since its invention and now averages 70 Å (30 to 40 Å) is routine for some SEMs). The large depth of field of this microscope, along with its resolution, makes it fall into an intermediate position between the light microscope and TEM.

Its imaging is somewhat similar to that of television (TV) where all focusing takes place prior to interaction of the beam with the specimen and the image points are addressed in time. This versatile instrument became popular among biologists in the 1970s. The first SEM was constructed by von Ardenne in 1938, but the microscope became commercially available only in 1965. The SEMs are manufactured by Amray (U.S.A.), Cambridge Instruments (England), Hitachi (Japan), International Scientific Instruments (made by Akashi, Japan), JEOL (Japan), Philips (The Netherlands), and by Zeiss (Germany). Attachments for SEM are available for newer models of commercial TEMs (*see* Fig. 5-1).

The SEM uses a small focused electron beam, 100Å or less in diameter, called *probe* that is formed by the demagnifying action of several lenses. The probe scans across the surface of the specimen in a predetermined raster pattern. Upon entering the specimen the primary beam interaction volume spreads into a teardrop shape (Fig. 17-1b). A number of signals at or near the specimen surface that include secondary and back scattered electrons are produced. These two electrons along with the electrons absorbed by the specimen (specimen current) may be used for SEM imaging. However, low-energy secondary electrons ($<50V$) that are more efficiently collected by a positively biased detector are most commonly used for SEM images. Back scattered electrons also enter the detector. They alone can be collected by reversing or eliminating the bias. The absorbed electrons (Fig. 17-1a) are usually not important for work with biologic SEM. The collected secondary electrons are then amplified and converted into an electronic signal, which is displayed on a cathode ray tube (CRT), essentially a TV screen. The probe spot scanning of the specimen surface is synchronized with that of the CRT so that one-to-one correspondence is obtained between the points on the specimens and the CRT.

The Design of the SEM

There are two basic components in an SEM. The first contains the microscope column and the second consists of electron collection, signal amplification, and image display system. A commercial SEM and its basic components are shown in Figure 17-2.

The SEM Column

The column of an SEM is schematically presented in Figure 17-3.

THE ELECTRON GUN: This is identical to that of a TEM, and a hairpin tungsten filament that emits electrons by thermionic emission is most commonly used (*see* Chap. 5). The accelerating voltage in SEMs varies from 1 to 50 KV. Most biologists operate at 10 to 20 KV, depending on the specimen. Recently, a high-brightness point source gun containing a fine-tipped lanthanum hexaboride in the form of a 1 mm solid rod has been used as the cathode in some SEMs. It produces a much brighter source at 1600° C than the tungsten filament, but the filament life is usually short when operated at a conventional vacuum of 10^{-4} to 10^{-6} torr. A yacuum near 5×10^{-7} torr in the gun chamber decreases contamination and increases the filament life considerably. Some SEMs use ion-pumped gun chambers (*see* in this Chap.); field emission or cold cathod type, extremely bright electron guns, are also available for SEMs (*see* in this Chap.).

ELECTRON LENSES: Electromagnetic lenses used in conventional TEMs are also used in SEMs. Together, they serve to demagnify the electron source to form the final *probe spot* that scans the specimen surface. Three lenses with varying aperture sizes are commonly used for this purpose. They are called first condenser, second condenser, and final condenser (also called final or objective lens), respectively. The final diminishing lens is used to focus the beam on the specimen, and its demagnification determines the diameter of the beam or probe size. The specimen resolution is primarily determined by the diameter of the probe. One or more apertures are used depending on the design of the SEM. The final lens is equipped with a stigmator to correct the astigmatism that is critical in determining the final image quality.

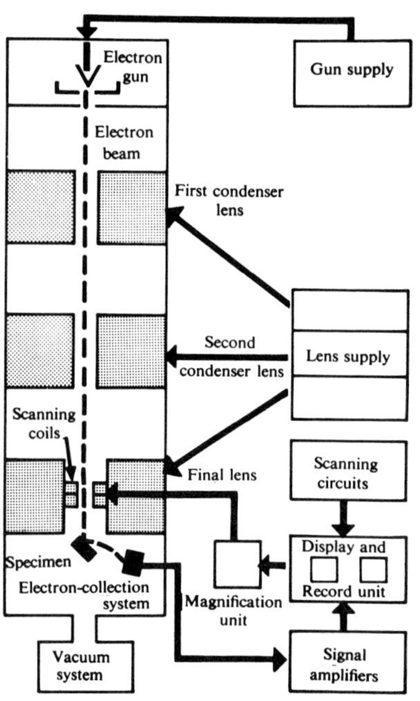

Figure 17-2. The scanning electron microscope (SEM) and its components. (a) The ISI-60 scanning electron microscope. (b) The block diagram of an SEM.

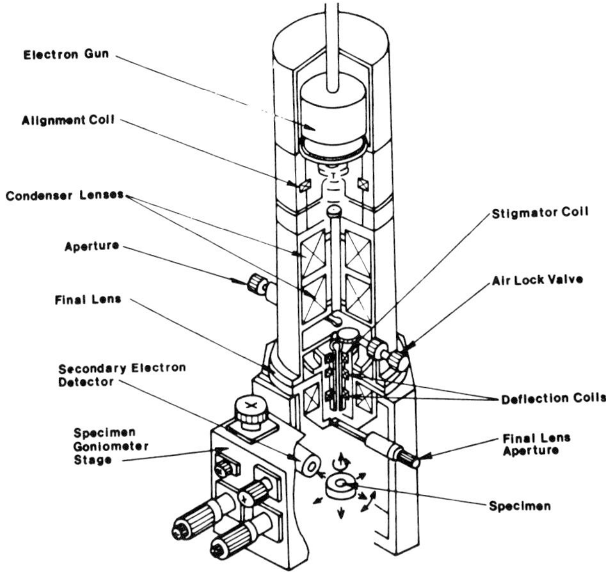

Figure 17-3. The schematic diagram of an SEM column. Courtesy of Hitachi Ltd.

THE SPECIMEN STAGE: A goniometer specimen stage with which one can orient the specimen in any direction under the beam is employed. It is located below the final lens, and two basic designs are used by various manufacturers. In the pre-pump chamber design, most of the mechanical parts of the stage assembly are inside the vacuum chamber, and an airlock system is used to insert the specimen. For the drawer type of chamber, the entire chamber is vented and the stage is removed like a drawer.

THE VACUUM SYSTEM: This is similar to that of the TEM. A mechanical pump and an oil diffusion pump hooked in a series are commonly used to obtain a vacuum of 10^{-4} to 10^{-6} torr. Turbo-molecular, cryopumps, or ion pumps may also be employed in some SEMs.

Image Formation

In the SEM, as the name indicates, the specimen is traversed by a focused probe spot.

Scanning

The movement of the probe across the specimen surface is usually achieved by two sets of scanning (raster) coils in the microscope column controlled by a scan generator. The primary beam is electromagnetically deflected across a given area of the specimen as shown in Figure 17-4. The probing beam starts, for example, at point A in the left and moves across to a point A′ in the right, then from B to B′ and so forth, thus building the image of the specimen line by line. After reaching the right-hand side, the beam is deflected back to the left and again starts on the next line. A probe spot-scanning in a raster of 600 to 1,000 lines is usually employed. The scanning of the probing beam is in synchrony with that of the CRT to yield a point-to-point translation. The two scannings start at the top at the same instant, sweep through the same number of lines, and end up at the bottom of their arrays at the same instant.

Signal Collection

Low-energy secondary electrons emitted at the specimen surface and high-energy back scattered electrons emitted from a greater depth can be collected for image formation in the SEM. Although most SEMs are equipped for collecting either of these signals, secondary electrons are more commonly used for biologic specimen imaging in the SEM. Some secondary electrons may be absorbed by the specimens, but those emitted near the surface after deflection along curved trajectories can escape and are collectable. The

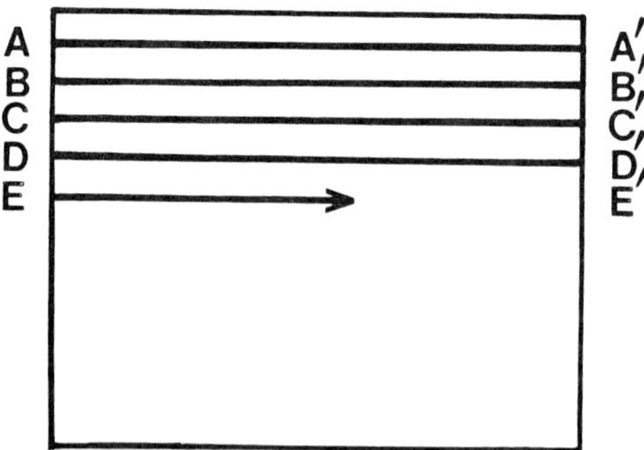

Figure 17-4. Scanning of specimen surface area.

number of these electrons is dependent on the nature of the surface and its angle to the incident beam (the specimen is usually placed at an angle to the beam).

The collector in the column consists of a cylindrical metal box with a copper gauze window. A phosphor-coated plastic scintillator is placed inside the collector. The scintillator is connected to the photomultiplier tube placed outside the column by a light pipe made of Plexiglas or polished quartz. The positively biased collector accelerates the secondary electrons toward the scintillator through a potential difference of a few hundred to a few thousand volts. Only a small number of back scattered electrons enter the detector.

When struck by secondary electrons, the phosphor of the scintillator immediately produces a flash of light (photon) that is short lived. The photons produced by the scintillator are then transported through a light pipe from the evacuated microscope column to the photomultiplier.

Signal Amplification

The photons carried by the light pipe are converted to an amplified electronic signal by a photocathode and photomultiplier tube placed outside the column. This signal is used to modulate the intensity of the CRT beam as it sweeps across the face of the tube in synchrony with the path of the beam over the specimen.

Image Display and Recording

The final magnified image is formed on a CRT, and SEMs generally have two CRTs, one for direct viewing and the other for photographic recording of the image. The brightness (amplitude contrast) on the screen is proportional to the number of secondary electrons emitted from the specimen. The CRT is in many ways similar to the picture tube of a TV. The deflection coils of the CRT are coupled to the scanning coils of the microscope, which produces a one-to-one representation of the area scanned. When all points of the specimen are scanned in sequence in a regular array, the points follow one another rapidly. Thus, the image of each point becomes an image of a line, and the line in turn can move down the screen so rapidly that a complete picture is visible to the eye (similar to a TV imaging).

The visual display CRT has a long persistent phosphor screen, which facilitates focusing and selection of specimen area to imaging. The CRT for photography has a short persistent phosphor screen to avoid fogging of the photographic film caused by afterglow. A slow scan rate, e.g. 10 to 40 seconds or longer per frame, is used for photograpic recording. A single long-interval scan rather than a multiple scan is generally used for making the photograph sharp. However, a much faster scan rate is required for direct viewing of the image so that it can be integrated by the brain. A scan rate ranging from $^{1}/_{30}$th of a second to a few hundredths of seconds can be used.

All SEMs are equipped to take pictures with Polaroid films, and positive-negative Polaroid films are recommended for making permanent negatives of the image. *Color photography* is possible, and some workers have used the signal strength for determining the color of each point of the image. A given signal intensity is converted to a particular color in such a *color modulation system.* Once the operator assigns a particular color to a given signal intensity, the color pattern of the image is determined by the signal generated from the specimen surface.

Contrast

Although secondary electrons are the principal source of contrast in an SEM, the contrast mechanism in the microscope appears to be complex. Factors such as specimen composition, specimen topography, and the angle of incidence of the electron beam influence the contrast. The number of secondary electrons produced depends partly on the density of the specimen, and the higher the atomic number, the greater the number of electrons. The conductive coating of the biologic specimen may also influence the emission of these electrons.

The topography of the specimen affects the detection of electrons. Since the detector is positioned to one side of the specimen, the electrons from that specimen area are collected. However, many electrons emitted from the specimen area facing away from the detector are blocked. This gives a shadowing effect, and the three-dimensional surface topography seen in the SEM image is the result of the distribution of light and dark areas. This is largely because more secondary electrons are collected when the beam strikes sharply curved areas than when it collides with a flat surface. This effect is accentuated by biologic specimens that are generally heavily contoured. If the specimen surface is smooth, tilting it at an angle to the probe beam provides the desired effect. The image contrast can also be controlled by regulating the voltage in the photomultiplier tube.

Magnification and Resolution

Magnification is the ratio of the length of a line on the CRT to a comparable line scanned on the specimen. It is controlled by the raster coils, and a change in magnification alters the size of the specimen area scanned. Thus, magnification can often be varied over a wide range without refocusing. A change of magnification also occurs when the accelerating voltage is changed. The useful range of SEM magnification varies from 10 × to 50,000 ×. The resolution is primarily limited by the probe spot size, but it may also be affected by the signal-to-noise ratio in the final image. The slower the scan rate the better is this ratio. A SEM micrograph may require a total scan time of one minute or longer.

Accessories such as gamma control, autofocusing control, Y-modulation, raster rotation, tilt correction, dual magnification, independent video channels, automatic data display, and automatic exposure are now becoming standard for modern SEMs.

Specimen Preparation

A relatively short specimen preparation time is one of the advantages of the SEM. Metals or conducting specimens can be examined immediately after their mounting on a specimen stub. Biologic specimens such as tissues, cells, and organisms, however, require careful preparation, and the preparative methods may be complex for some specimens. Even then, the time and labor involved in such preparations are considerably less than required for ultrathin sectioning of biologic specimens. Specimen preparation is often the factor that limits the quality of a scanning electron micrograph. Adequate and careful specimen preparations are, therefore, necessary to take advantage of the high resolution of the modern SEMs.

pressure of 31° C and 1072 lb/sq inch, respectively), although some workers have used several types of fluorocarbons. In the original technique of Anderson (1951), the specimen was first dehydrated in graded strengths of alcohol and then placed in amyl acetate before critical point drying in liquid CO_2. Recently, the use of amyl acetate has been eliminated.

The specimen in absolute alcohol is transferred to a critical pont drying apparatus (Fig. 17-5) that has a transparent window made of sapphire, quartz, or certified tested glass for observing the exchange of fluids. Pressurized liquid CO_2 is allowed to enter and fill the vessel for exchanging the intermediate fluid (alcohol). During this time, the exhaust valve of the chamber is slightly opened to allow the fluid to flush through the vessel and sample for carrying away the alcohol and air. When the alcohol is no longer detectable in the exhaust, the chamber is resealed at both the exhaust and inlet valves. The temperature of the vessel is then gradually increased, during which time the pressure of liquid CO_2 also increases. When liquid CO_2 passes through the critical point, its gas can be bled off gradually while the temperature of the vessel is kept at 31° C; the specimen is then dry.

Specimen holders for critical point drying are available, but some specimens may require modified specimen holders. Adequate safety precautions should be taken in using the critical point drying apparatus. The rupture disk in this apparatus may blow out, and it should face away from the person using the apparatus. A fume hood is recommended for avoiding the possible toxic effects of liquid CO_2 and fluorocarbons.

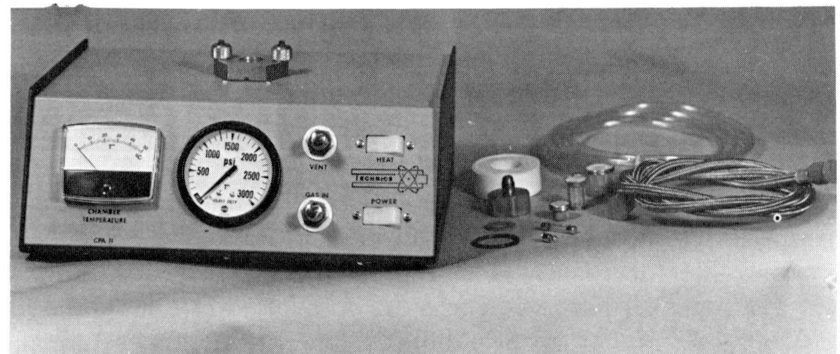

Figure 17-5. A critical point dryer. Courtesy of Technics East, Inc.

Specimen Mounting

The sample is usually mounted on aluminum or brass stubs of various sizes, and most SEMs accommodate more than one size. Carbon stubs may also be used if x-ray microanalysis of the sample is being made. A variety of electrically conductive adhesive materials, e.g. silver paints, silver pastes, conductive epoxies, and double stick tapes, have been used to glue the specimen to the stub. A double stick tape is quite convenient. Specimen orientation is important, and a dissecting microscope may be used for mounting small specimens.

Specimen Coating and Coating Methods

It is desirable to maintain the electric potential of the specimen at ground level. Biologic materials (insulated materials) do not conduct electricity and heat during their examination in the SEM, and such problems as *charging* by the beam and *beam damage* are encountered with these samples. Charging imparts a negative charge to the surface, which cannot be dissipated. They can be coated with a thin layer of conductive material (100 to 1,000 Å thick) to overcome these difficulties. Electrical connection made between the film and the specimen stage helps to carry away the charge. An improved collection of secondary electrons may also result due to this coating. Beam and heat damages can occur at high beam current, especially during x-ray microanalysis.

Among a variety of metals used for coating, gold-palladium is the most popular coating metal for biologic specimens. Carbon coating is preferable for x-ray microanalysis because it does not produce interfering x-rays. Antistatic sprays such as Duron tend to cover small surface details and are unsuitable for high-resolution work.

Gold-palladium is evaporated onto the specimen mounted on the stub by the resistance heating method in a vacuum evaporator. Rotary shadowing (*see* Chap. 11) is necessary for depositing a continuous and uniform layer of the metal on all surfaces and for avoiding shadows.

Recently, *sputter coating* has become popular because it is cheaper and is rapidly replacing rotary shadowing. Several types of sputter coating instruments that include ion beam sputter coater, radio frequency sputter coaters, and direct current sputter coaters of diode or triode design are used for this purpose. Direct current sputter coaters (Fig. 17-6) are most widely used.

A relatively low vacuum (10^{-2} torr) is used in the sputtering chamber, and a mechanical pump is adequate for this purpose. Sputtering occurs when a target metal is bombarded by energetic particles. The gas molecules in the chamber are air, inert argon, or nitrogen. When these molecules are ionized, excited to a rapidly moving state by glow discharge, they bombard the commonly used target metal gold-palladium, and they dislodge the atoms from the target material. The dislodged gold-palladium is then uniformly deposited on the specimen surface regardless of the orientation.

An electric field is established between the cathode (generally the target material) and the anode (generally the specimen). The gas molecules ionize and produce large positive ions and free electrons in the electric field near the cathode. These positive ions then bombard the negatively charged target material and dislodge the

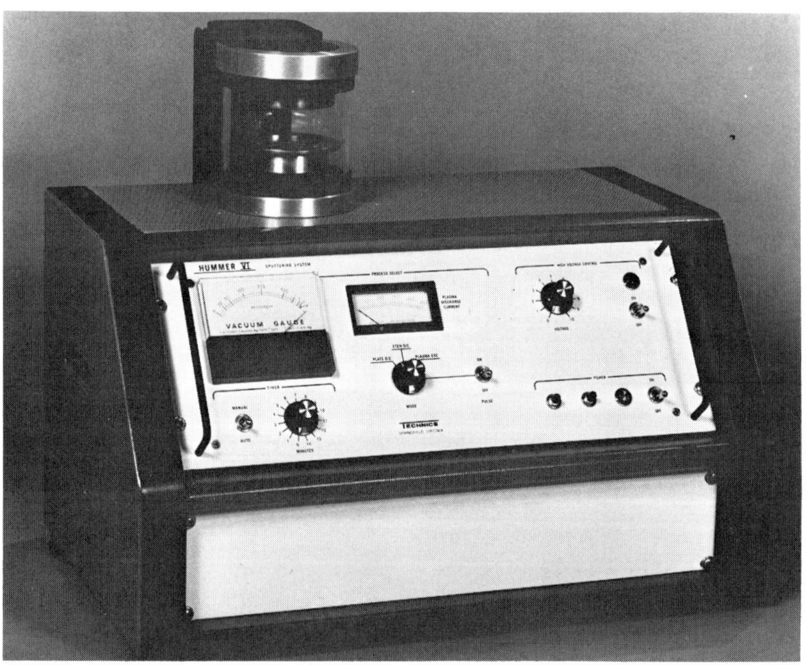

Figure 17-6. A sputter coater. Courtesy of Technics East, Inc.

target particles. Sputtering of the target particles by a low vacuum and excited gas molecules is called *plasma sputtering*; the excited state occurring between the cathode and anode is called *plasma*. It is characterized by *glow discharge*, and a dazzling lavender glow of air and argon is seen.

Examination in the SEM

After metal coating, the specimen is examined in the SEM. The astigmatism should be corrected, and a low beam current and a low accelerating voltage should be used as much as possible. A number of scanning micrographs using secondary electrons are illustrated in Figure 17-7. A comparison of back scattered and secondary electron images is presented in Figure 17-8. Stereopair SEM imaging is also useful for some specimens.

Other Specimen Preparation Techniques

In addition to examining the whole specimen or tissue surface in the SEM, various other methods have been used. These include sectioned tissues before or after embedding in paraffin or epoxies, freeze-fractured and freeze-etched specimens, and replicas.

Ion Beam Etching (Sputtering)

This technique has been employed to reveal underlying surfaces of some biologic specimens. A beam of ions such as argon is used to blast away the surface of the specimen to reveal the underlying material. The principle of the sputtering technique has been previously described (*see* in this Chap.).

Cell Surface Labeling

Immunologic methods similar to those used in the light microscope and TEM may also be applied to the SEM. Latex particles can be covalently or noncovalently coupled to the antibody, and this can be used to localize antigen on the cell surface. Other antibody markers include silica spheres, tobacco mosaic virus, and T_4 bacteriophage. Concanavalin A-hemocyanin and colloidal gold have also been employed as markers.

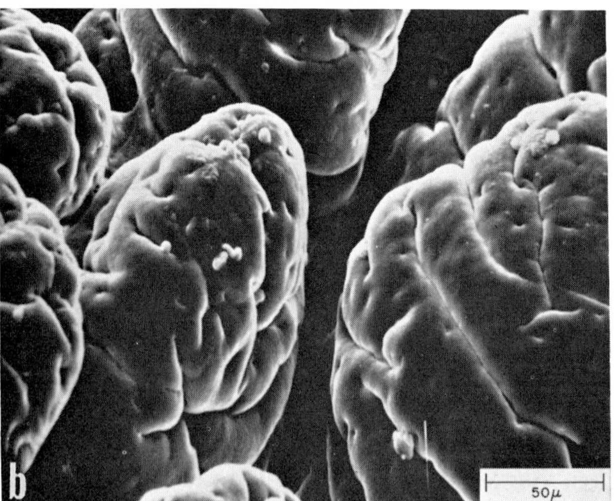

Figure 17-7. Scanning electron micrographs. (a) *Aspergillus spp.* (b) Villi on the mucous membrane of human small intestine.

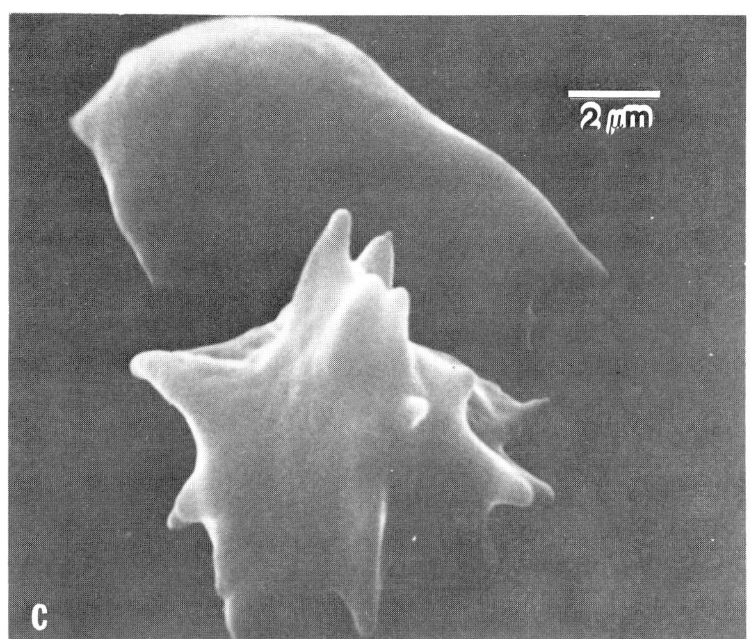

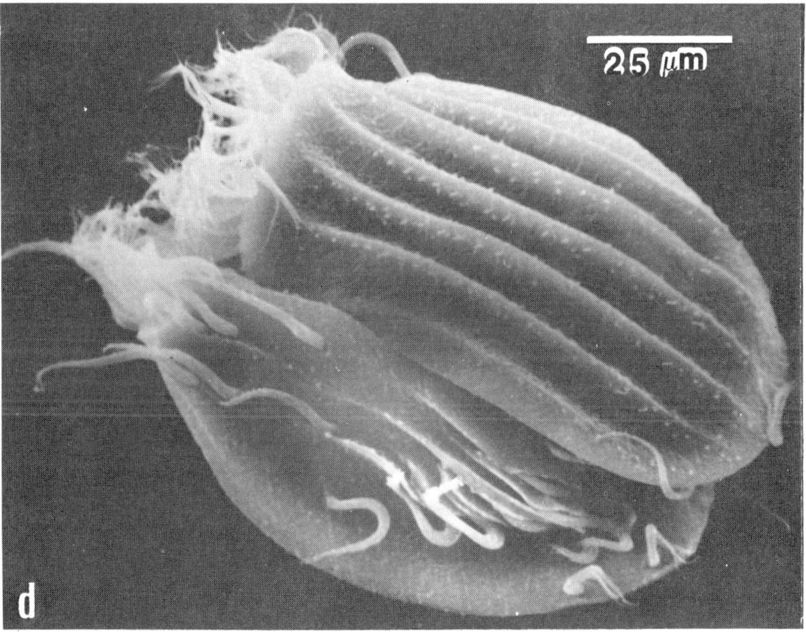

Figure 17-7. (c) Blood cells from a sickle-cell anemia. (d) A conjugating pair of *Euplotes*.

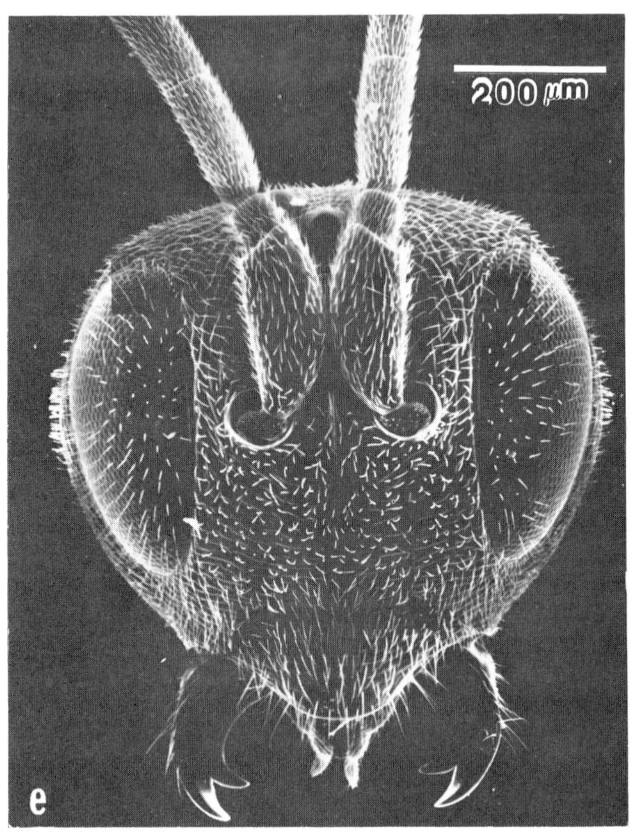

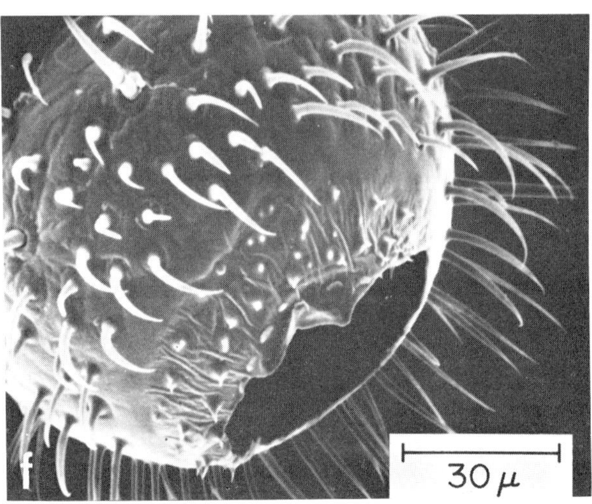

Figure 17-7. (e) Wasp head. (f) A segment of the antenna of a termite. a–f Courtesy of Electron Microscope Central Facility, University of Maryland.

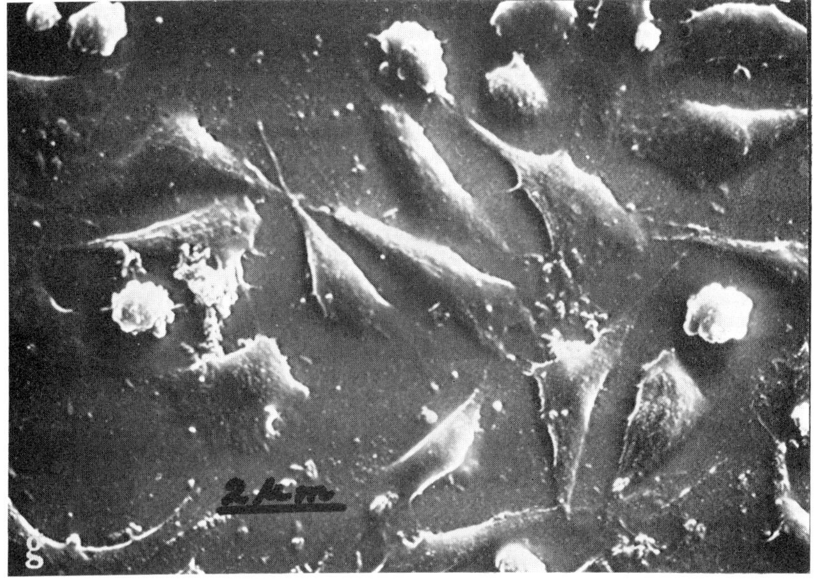

Figure 17-7. (g) Cultured bovine embryonic kidney cells.

Living Specimens

A few specimens, such as some insects that can tolerate dehydration, vacuum, and radiation, have been shown to recover after their examination in the SEM. They require no fixation, dehydration, or coating.

The SEM has become a valuable research tool for the biologist as evidenced by the large number of these instruments now in use. As the resolving power of this instrument improves in the future, it will even be more useful to biologists.

SCANNING TRANSMISSION ELECTRON MICROSCOPY

Since the demonstration of heavy atom images by Crewe et al. in 1970, virtually everyone involved in electron microscopy is aware of the capabilities of scanning transmission electron microscopy. The scanning transmission electron microscope (STEM) is the most sophisticated instrument in the family of electron microscopes. Its resolving power is now comparable to that of the TEM (2 to 3Å), and it is expected to be better in the future.

Figure 17-8. Back scattered and secondary SEM micrographs. (a) Back scattered electron SEM image of a specimen grid with a thin section. (b) Secondary electron SEM image of the same specimen.

Instrumentation

The STEM is similar in concept to the SEM except that thin specimens are used and the electrons transmitted by the specimen are used to modulate a CRT, thus producing a transmission image by a scanning mode. It was A.V. Crewe and his group in Chicago who employed a field emission source, an ultrahigh vacuum with ion pumps, a suitable detector system, electron velocity spectrometers to construct the first high-resolution STEM in the late 1960s. A commercial model high-resolution STEM and its components in a block diagram are shown in Figure 17-9.

Field Emission Gun

The electron probe spot is formed by a cold cathode, point source triode type field emission gun, which is focused on the specimen by one or two demagnifying electromagnetic lenses (two lenses are now commonly used). The electron emitter is a very sharp tungsten point electrolytically etched to a radius of a few hundred Å (Fig. 17-10). An electrostatic field formed by an electrode placed close to the tip drags the electrons away from the fine point tip. A moderate high voltage, 1 to 5 KV applied to the tip, is sufficient to produce an extremely large electric field at the point for electron emission.

The brightness of this source is $10^5 \times$ greater than the conventional thermoionic source, but $10^3 \times$ as small. The field emission source is also highly coherent and virtually free of chromatic aberration. The life of the filament is extremely long—months and sometimes up to two years (compared to hours with a hot tungsten filament). When the tip becomes contaminated, *flashing* the tip by passing a current to raise its temperature removes the contamination. The electrons are accelerated through 50 to 100 KV.

Ion Pump

The field emission source requires an ultrahigh vacuum (10^{-9} to 10^{-12} torr). Therefore, heaters for outgassing the metal parts in the gun chamber and ion pumps that require no pumping fluids and moving parts are used. This ultrahigh vacuum is necessary only in the immediate neighborhood of the tip and only otherwise around the specimen.

In the ion pump, a layer of highly reactive metal such as titanium

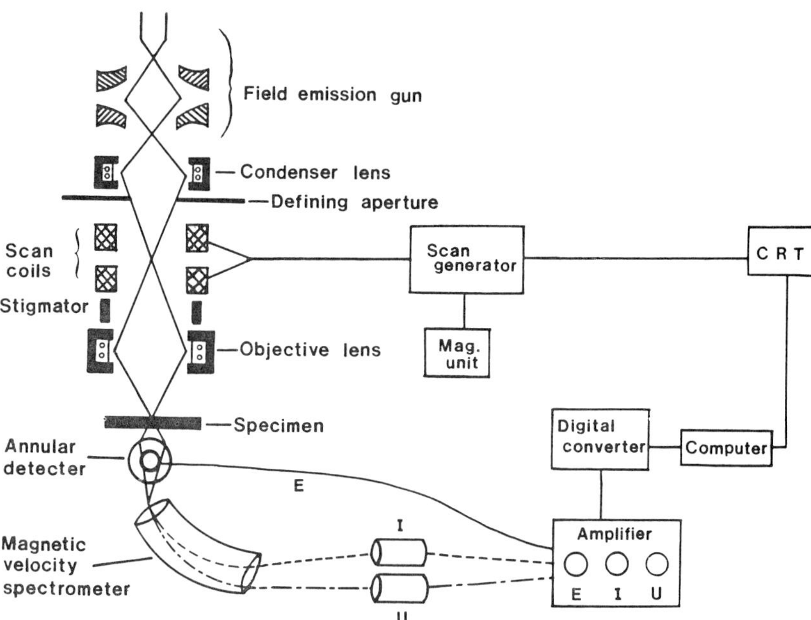

Figure 17-9. The scanning transmission electron microscope (STEM). (a) A commercial high-resolution STEM. VG microscopes model HB-501. Courtesy of V.G. Microscopes. (b) A block diagram showing the components of a high-resolution STEM. E = elastically scattered electrons; I = inelastically scattered electrons; U = unscattered electrons; CRT = cathode ray tube.

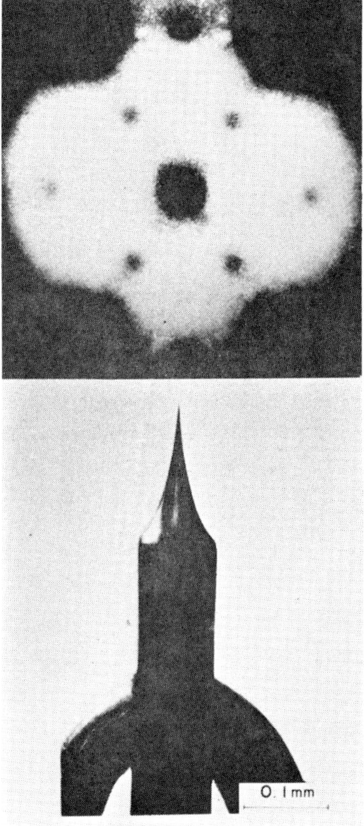

Figure 17-10. The field emission tip (at the top) and its emission pattern (at the bottom). Courtesy of Hitachi, Ltd.

or zirconium is continuously evaporated from a heated metal filament or refractory surface onto an inert cathode placed in the emission chamber. Initial ionization of residual gas molecules can be achieved by adding a third electrode held at about 1 KV. The gas molecules then combine with highly reactive titanium and are removed from the chamber.

Lenses and Scanning Coils

A two-lens (conventional TEM lenses) demagnifying system is used to form an intensely bright probe spot of 2 to 5 Å diameter. The condenser lens is outside the ultrahigh vacuum as are the scanning coils and stigmator. The scanning coils are of a double deflection type, and a stigmator is placed near the back focal plane of the objective to make the probe spot circular. The electronics of the STEM are similar to those of the SEM, but the scan generator in the STEM is digital. This allows the microscope to be interfaced with a computer for instrument control or for image storage and processing.

Detector System and Image Display

The STEM is extremely versatile in that various transmitted signals virtually free of chromatic aberration can be collected and displayed on various CRTs. Electrons transmitted by the specimen are collected by a detector arrangement. An annular detector is used to collect the elastically scattered electrons (*see* Chap. 4) to form a high-resolution dark-field image. The signal that passes through the annular detector is the sum of the unscattered and inelastically scattered electrons (*see* Chap. 4). This signal can be passed through a *velocity filter* (*electron energy analyzer, energy spectrometer*), a sector-shaped magnetic field, and unscattered and inelastically scattered electrons can then be separated (filtered). When detected separately, the unscattered electrons produce the conventional bright-field image and the inelastically scattered electrons provide a dark-field image. The energy analyzer may also be used to produce the energy loss spectra to serve as a basis for chemical analysis of small specimen regions.

The three signals can be separately amplified and used to modulate the brightness of three separate synchronous CRTs for visual observation and for photographic recording. The signals can be digitized and then processed by a computer. The most important aspect of the STEM is that the image signals, contrast, and intensity can be readily manipulated by electronic means. Images from different signals can be added, subtracted, multiplied, or divided at will. It is possible to clean up the signal and remove random noise. Contrast can be increased, decreased, or reversed and the signals recorded on a magnetic tape for later analysis.

Magnification and Resolution

Overall magnifications in the order of millions can be obtained without imaging lenses. The ultimate resolution of the STEM is limited by the probe beam diameter and the spherical aberration in the objective lens. As the magnification is changed there are no changes in focus, brightness of the image on the CRT, or image rotation.

Capabilities of the STEM

The great potential of the STEM in biology has been amply demonstrated. Single heavy atoms were imaged by using this microscope by Crewe et al. in 1970. A 20Å-carbon film baked at 2700° C in argon atmosphere was used to place a long-chain polymer of benzenetetracarboxylic acid linked by uranium or thorium atoms spaced 10Å apart. Using the signal formed by the elastically scattered electrons, the atoms were visible, and the spots disappeared when inelastically scattered electrons were used for imaging. The ratio of inelastically scattered electrons to elastically scattered electrons and the visibility factors were close to the calculated ones within reasonable experimental errors. The signals from the atoms were also 68% brighter than that from the carbon film. Single silver atoms that are labile under electron bombardment have also been seen, and movies of the atoms have also been taken.

The high-resolution and high-contrast dark-field signals formed by elastically scattered electrons are the real advantages of the STEM. They are most useful for examining unstained biologic specimens in the STEM. A variety of specimens including herpes simplex virus capsids and T_4 bacteriophage have been examined, and their basic structure is the same as seen in negatively stained preparations. A much larger signal for a given incident electron beam is obtained than the conventional TEM dark-field imaging. Thus, the radiation damage to the specimen is much more reduced than that of the TEM, and the signal-to-noise ratio is improved.

By using this signal, helicity and strandedness of DNA, periodicity of tail structures of T_4 bacteriophage, and tobacco mosaic virus have been demonstrated. Since there are no imaging lenses, there is no chromatic aberration, and thicker specimens can be examined. The resolution of image using inelastically scattered electrons, however, is 10 to 15Å (vs. 2 to 3Å for elastically scattered electrons). This is good enough to collect elastic and inelastic signals simultaneously from a specimen area and calculate the average atomic number and the approximate mass. A Z-contrast can also be produced by the ratio of the collected elastically and inelastically scattered electrons, and this contrast has been used for examining unstained thin sections. Electron micrographs of unstained thin sections examined in a commercial STEM are illustrated in Figure 17-11.

Although it is impossible to prevent the radiation damage, this damage can be significantly reduced especially when the specimen is kept at liquid helium temperature. Furthermore, an image can be formed after a single passage of the probe beam. The sequencing of DNA has not yet been achieved with the STEM because the heavy atom stain does not remain fixed in space. The STEM is generally not as efficient as TEM in obtaining phase contrast or diffraction

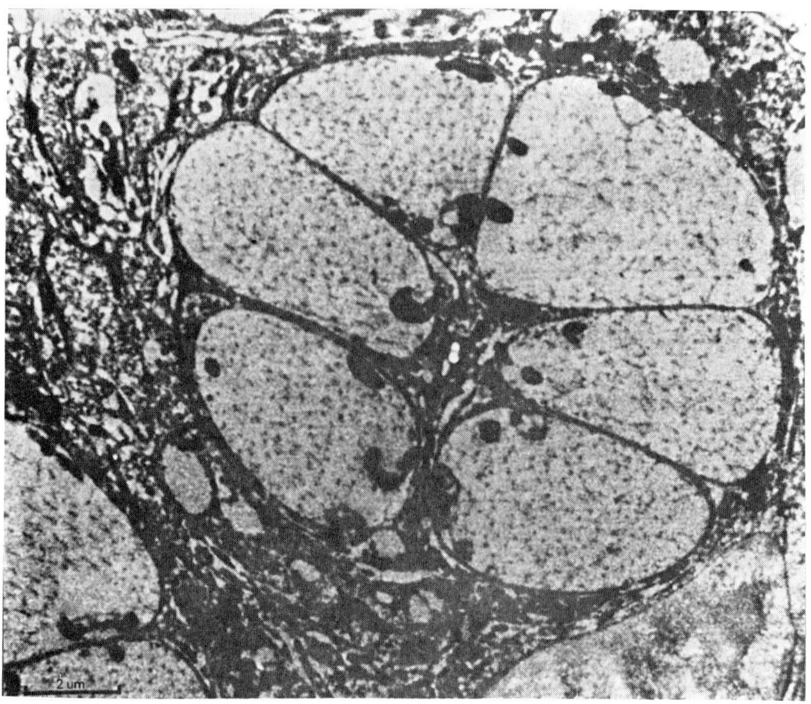

Figure 17-11. The STEM electron micrographs of unstained thin sections taken in VG Microscopes model HB-5. Prepared by the conventional technique but without postfixation with OsO_4 and poststaining of thin sections. Images were recorded using the mixed signal mode unique to the STEM, (bright field − dark field)/(bright field + dark field). (a) The optical system of blowfly *Calliphora erythrocephala*. The first optic neuropile shows photoreceptor axons.

information. Contamination is the main problem in a high-resolution STEM, and the specimen itself can be the main source of contamination. However, scrupulous clean handling of the specimen and use of well-designed ion pumps can virtually eliminate this effect.

There are approximately twenty-five high-resolution STEMs in the world. Of these, only five have resolving power high enough to obtain a single atom picture (for which a point resolving power of $\leq$ 4Å is required). Two of these are in Chicago, and one each at Johns Hopkins University, Brookhaven, and the Hitachi microscope in Japan. A high-resolution STEM (guaranteed lattice resolution of 2.04Å) with a field emission source, electron spectrometer, and image analysis facilities is now available commercially (Fig. 17-1a) from VG Microscopes, Model HB501, England. A prototype of this instrument has been built by the AEI (England). Major EM

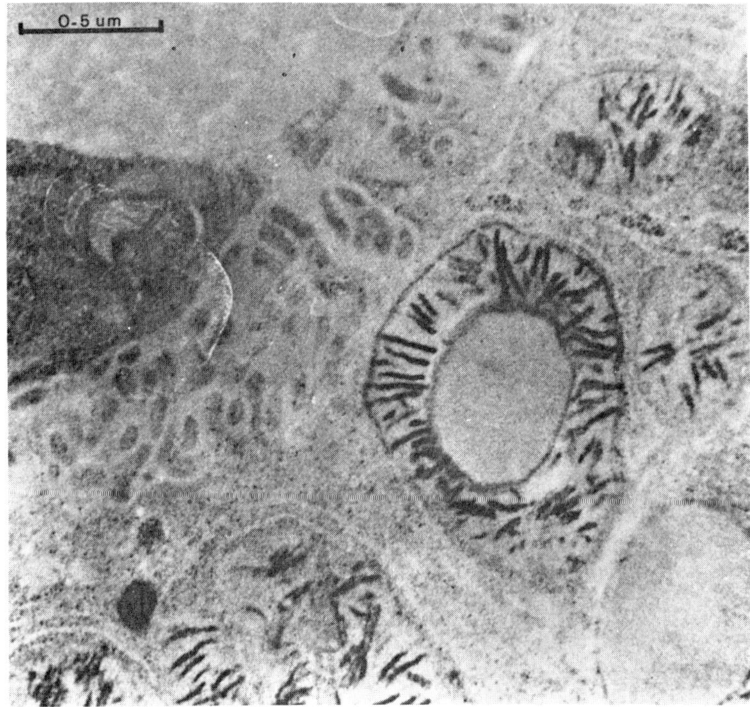

Figure 17-11. (b) Hepatocytes in rat liver showing cytochrome oxidase activity in mitochondrial cristae. Courtesy of V.G. Microscopes.

manufacturers are now heavily engaged in developing high-resolution STEMs. One reason for the slow development of this highly complex instrument, of course, is the high cost.

STEM Attachments for TEM

Attachments for STEM are now available for many modern new model high-resolution TEMs. Figure 5-1 shows such a TEM with SEM, STEM, and x-ray microanalyzer attachments, a total analytic system. In the absence of a field emission gun, an electron spectrometer, and image analyzer, the resolution of these instruments would always be comparatively poor (10 to 25Å). A third condenser lens is used in these instruments to make a probe spot of 100Å, and the objective lens is switched off when the STEM mode is used. Beam deflection coils are used to form a raster on the specimen in synchrony with the CRT. A Faraday cage is used as the detector, which is placed under the objective lens to collect the transmitted electrons.

The signal is amplified and modulates the brightness of the CRT, and a low-resolution bright-field image is then visible. The important advantages of these STEM attachments are the electronically amplified and reversible image contrast. Thus, a high-contrast image can be obtained from specimens that would otherwise have little contrast in the TEM. The damage to the specimen is also reduced because the probe beam interacts with a particular specimen area only once per scan.

Crewe and co-workers in Chicago are actively engaged in building a 1 MV STEM that is expected to approach the atomic resolution (1Å), and thicker specimens can then be examined. The group is also working on a spherical aberration corrector for the objective lens, and attempts are being made to improve the photographic recording (possibly color photography). It appears that the STEM may prove to be the greatest development since the EM was first built in 1931.

X-RAY MICROANALYSIS

Characteristic x-rays are emitted (Fig. 17-1a) when the high-energy electron beam interacts with the specimen (*see* Chap. 4). An electron microprobe instrument, whether it is TEM, SEM, or

STEM, can incorporate an x-ray detector and analyzer and a display unit. Many SEMs are now equipped with x-ray microanalysis attachments; a TEM with all the attachments is illustrated in Figure 5-1. X-ray microanalysis is finding increased usage by biologists, but it is more widely used by material scientists, metallurgists, and geologists.

The major problems in x-ray microanalysis are the detection, identification, and quantification of x-rays emitted from the specimen. X-rays produced by the sample have wavelengths and energies that are characteristic of the elements in the specimen. Therefore, these x-rays can be detected by the *wavelength dispersive spectrometer* (WDS) or *energy dispersive spectrometer* (EDS). The WDS can detect all elements with atomic number (Z) $\geqslant$ 4 (Berrylium). The EDS usually operates with $Z \geqslant 11$ (sodium) with Berrylium window and $Z \geqslant 6$ (carbon) without the window. The WDS is time-consuming, and it draws the peaks on a chart; EDS is more sensitive, and the spectrum can be displayed on the CRT. The x-ray microanalyzer can operate either in a probe unit or scan mode.

X-ray analysis has given information on the distribution of calcium in the bone and tooth, localization of iron in the haemopoetic system, and incorporation of drugs containing metals in tissues. X-ray microanalysis plays an important role in forensic science. The technique has been used for detecting lead levels in infants' teeth, heavy metals in tissues, and asbestos fibers in lungs.

SELECTED BIBLIOGRAPHY

Crewe, A.V., Langmore, J., Wall, J., and Beer, M.: Single atom contrast in a scanning microscope. *Proc EMSA, 28th Annual Meeting*, 250, 1970.

Crewe, A.V.: Scanning transmission electron microscopy. *J Microscopy, 100*: 247, 1974.

Crewe, A.V.: High resolution scanning microscopy—what next? *Proc EMSA, 33rd Annual Meeting*, 18, 1975.

Crewe, A.V.: Is there a future for the STEM? *9th International Congress in Electron Microscopy*, vol. III. Toronto, 197, 1978.

Crewe, A.V.: Something old, something new for the STEM. *Proc EMSA, 37th Annual Meeting*, 560, 1979.

Drummond, I.W.: Scanning transmission electron microscopy. *Am Lab, 8*:83, 1976.

Goldstein, J.I., Newbury, D.E., Echlin, P., Joy, D.C., Fiori, C., and Lifshin, E.: *Scanning Electron Microscopy and X-Ray Microanalysis*. New York, Plenum, 1981.

Types of Specimen Damage

Four types of specimen damage (*radiation damage*, *beam damage*) can occur in the TEM. These are radiation displacement damage, ionization damage, beam heating, and charging.

Radiation Displacement Damage

In this process, the incident electron transfers sufficient energy to a specimen atom so that the atom is displaced from its original position. This is what material scientists refer to as *radiation damage*. Radiation displacement damage probably does not occur or is insignificant in biologic specimens at 100 KV because it is thought that the threshold voltage for carbon atom displacement is about 120 KV. The threshold voltage, however, is dependent on the nature of chemical bonding of the specimen and radiation displacement injury becomes a serious problem with the high-voltage electron microscope (HVEM). Radiation displacement damage in metals is well documented, but data directly relating to biologic specimens are not available.

Ionization Damage

For this process, the electron-specimen interaction ejects an electron out of the specimen. Ionization of metallic specimens does not cause any structural change because the ejected electron is usually replaced by another electron flowing up from the earth due to the normal conducting path established by the specimen to the earth. If the metallic specimen is insulated from the earth, it becomes charged.

However, ionization frequently causes a permanent change in covalently bonded structures of biologic specimens because of the resulting electron rearrangement. This is often accompanied by an atomic rearrangement within the molecule. The maximum exposure to the electron before these changes occur is called *critical exposure*, which is dependent on the type of specimen and the energy of the incident electron beam. The probability of ionization damage in biologic specimens is far greater than the radiation displacement damage. Theoretically, it is inversely proportional to the square of the accelerating potential, and ionization damage is apparently reduced in the HVEM.

Heat Damage

Beam heating can cause injury to the specimen, and it varies with the accelerating potential. For a given specimen thickness, this heating decreases as the accelerating high voltage increases.

Charging

When electrons are transmitted through a biologic specimen, it can accumulate charge by loss of secondary electrons. This can destroy the specimen (*see* Chap. 4).

The Nature of Damage Process

The basic nature of radiation damage to biologic specimens in the TEM is not clear. The most important fact, however, is that the damage is an *electron dose effect*. This is proportional to the product of electron density and the radiation, provided there is a negligible heating of the specimen. To better understand the process of radiation damage, it is necessary to consider the electron scattering by the specimen (*see* Chap. 4).

Electrons can pass through the specimen without any interaction with the object, and these *unscattered electrons* cause no harm to the specimen. Incident electrons are *elastically scattered* through a wide angle without loss of energy after striking the atomic nuclei of the specimen. These electrons are useful for high-resolution imaging, produce high contrast, and cause negligible or no injury to the specimen. Electrons are *inelastically scattered* through a narrow angle with loss of energy by collision with the orbital electrons of the specimen atom. This process imparts energy to the specimen, causes excitation of the specimen atom, and is mainly responsible for causing molecular or chemical changes in the specimen. Since there are approximately 2 to 3× as many inelastic events, the specimen is inevitably harmed, and most of the damage is complete before even a small fraction of the image is produced.

An average of 30 to 40 V is transferred to the specimen by each inelastic event. This energy may be sufficient for causing ionization of the atom. It may also disrupt one or more interatomic bonds, create free, highly reactive radicals, or change the bonding configuration. Thermal denaturation, surface diffusion, and evaporation may also occur subsequent to the radiation damage. An appreciable mass of the sample is lost, and diffraction patterns of crystalline materials fade rapidly. Amino acids, proteins, and nucleic acids lose 40 to 90% of their mass during radiation damage, and this loss is proportional to the radiation dose. In a hydrated state, the formation of reactive OH^- radicals can have an appreciable effect on the radiation damage.

A few scattering events must occur before a specimen can be imaged in the TEM. An electron dose of 2,000 electrons/$Å^2$ is usually required for a single atom imaging in the TEM. The electron *damage dose* varies greatly with the specimen, and many dehydrated enzymes are inactivated at an electron dose of 10^{-3} electron/$Å^2$ at 100 KV. Most amino acids lose their crystallinity after exposure to 1 electron/$Å^2$, and a loss of mass occurs at 5 electrons/$Å^2$. The damage dose for killing *Bacillus subtilis* spores is 4×10^{-4} electrons/$Å^2$ at 100 KV, and observation of living specimens seems impossible. Although the electron dose for single atom imaging can be reduced to 30 electrons/$Å^2$ by the STEM (compared to 2,000 electrons/$Å^2$ in the TEM), and some specimens can be imaged with a dose of ≈ 10 electrons/$Å^2$, radiation damage is unavoidable in electron microscopy.

Measurements of Radiation Damage

Two basic approaches have been made to study radiation damage in the TEM. In the first method, the specimen is irradiated in the microscope and then removed to determine the effects of radiation. A number of different indicators including mass loss (weight measurement), changes in enzyme activity, elemental composition, loss of radioactivity, and change in infrared and visible spectra are then used to determine the extent of damage incurred by the specimen.

For a more direct approach, the specimen damage is evaluated *in situ* without removing the specimen from the microscope. Changes in the diffraction pattern of crystalline materials and changes in the characteristic energy loss spectra of electrons transmitted through the specimen can be measured by this method. A change in bright-field and dark-field image intensities, which indicates a loss of scattering power, may also be related to the loss of material.

Possibilities of Minimizing Radiation Damage

Although radiation damage in the EM is unavoidable, the situation is not as hopeless as it may seem. There are various methods by which the damage to the specimen can be minimized.

Minimal Exposure Technique

Since the electron dose determines the extent of damage, it is essential to irradiate the specimen during the photographic recording only. In this method, the focusing and astigmatism correction are done in one area of the specimen using a small illuminating beam spot on the screen. The illumination is then shifted to a different part of the specimen not previously illuminated at a high intensity, and the image is recorded immediately. The illumination shift is performed by a movement of the projector lens polepiece or by a beam deflector. The disadvantage of this technique is that the specimen is photographed blindly and the microscopist is not sure if the area was interesting until the exposure is complete. However, the method has given excellent results with negative staining (Fig. 18-1), and the difference between normal and minimal exposure has been remarkable. A relatively similar procedure is used in the high-resolution STEM.

The HVEM

Although ionization damage is reduced in the HVEM, the radiation displacement damage increases in this microscope. Therefore, there appears to be no substantial advantage in using the HVEM as far as radiation damage is concerned.

The STEM

Undoubtedly, the minimum electron dose necessary for specimen imaging in the STEM (Crewe type) is much smaller than the TEM. It

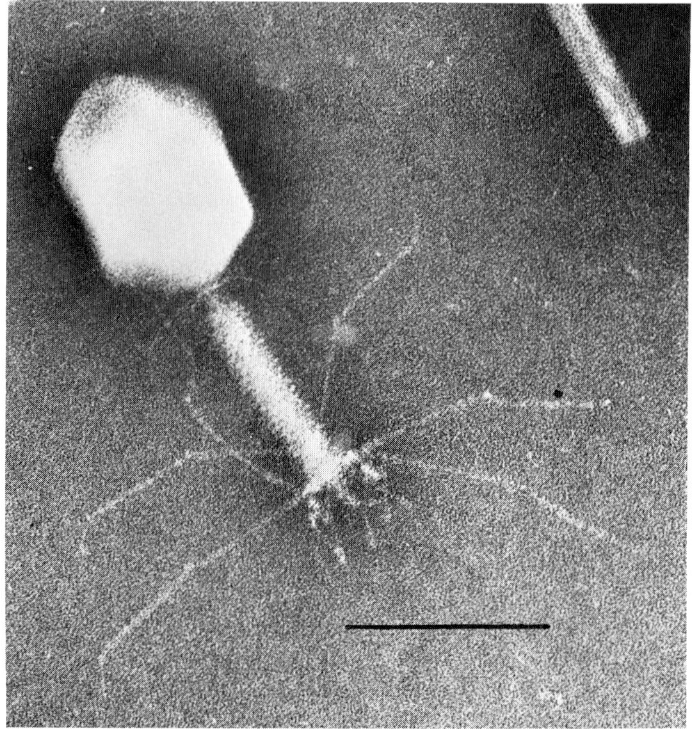

Figure 18-1. T_4 bacteriophage negatively stained with phosphotungstic acid and photographed with minimal beam exposure. Bar = 100 nm. Courtesy of Dr. R. C. Williams.

causes less damage to the specimen and can also extract a maximum amount of information for every incident electron. This microscope may be the answer to minimizing the radiation damage, but at present, there are only a handful of these high-resolution instruments.

Low Temperature

There are reports that liquid helium temperature (−269° C) does not seriously damage the enzymatic activity or crystallinity. Mass loss of biologic molecules is also dramatically reduced at this temperature.

Signal Averaging

Although a low radiation dose is needed for reducing the specimen damage, unfortunately the signal/noise ratio is substantially reduced, which may limit the effective resolution. This can be avoided by signal averaging of many noisy electron micrographs of many identical objects in identical orientation. The micrographs can then be superimposed and their signal averaged by various techniques including the computer. Remarkable results have been obtained with periodic structures, but it is more difficult for aperiodic objects.

It is obvious that the specimen damage is the single most important problem in biologic electron microscopy, and researchers are attempting to overcome this difficulty. The problem is not insuperable, and more research is needed on radiation damage of biologic specimens, particularly at low temperatures. Some multiatom stains apparently lower the radiation damage, and their development is clearly indicated.

IMAGE PROCESSING OF ELECTRON MICROGRAPHS

A transmission electron microscope (TEM) image essentially represents a two-dimensional projection of a three-dimensional object. The micrograph cannot represent the object faithfully because the specimen usually undergoes drastic preparative procedures to make it visible in the microscope. Artifacts are usually introduced in this process, and the specimen structure is often radically altered from its original state due to dehydration, distortion, and radiation damage. Instrumental aberrations, specimen drift, defocusing, multiple scattering, noise from the electron beam, substrate, and photographic emulsion also contribute to the imperfect imaging of the specimen.

Image processing techniques are used for examining and altering the information contained in the electron micrograph. The procedures can be used to obtain a clear image of the specimen by suppression of noise present in high-resolution electron micrographs. The term *image enhancement* refers to those techniques which improve the image quality and make pictures more interpretable. It is important that the micrographs be of high quality before such techniques can be applied for their enhancement. The resolution should be high, and an adequate coherent electron source should be used for imaging. Image processing of electron micrographs obtained by improper preparative methods and under poor instrumental performance conditions, for all practical purposes, is a waste of time.

Image processing of biologic specimens has increased rapidly since rotational superimposition and optical diffraction methods were introduced in 1963 and 1964, respectively. The technique has been most commonly used with negatively stained periodic objects. It is generally more difficult for aperiodic specimens, and image processing (digital computation) has been applied to a few of these objects only to a limited extent.

Image Processing Methods

Image processing commonly involves one or more of the reciprocal space or Fourier methods that include optical or digital diffraction and filtering and three-dimensional reconstruction. The processing consists of two steps in which the images are first selected and assessed, and an averaged image is then produced. In the first step, called *diffraction* or *Fourier transformation* step, a diffraction pattern of the image is produced. This pattern can separate the noise and signal present in the image so that they can be identified. The diffraction pattern of the image provides a useful measure of lattice dimensions, symmetry of periodic objects, specimen preservations, and aberrations introduced by the instrument. For the second step, *reconstruction* or *inverse Fourier transformation*, a mask is used to remove the noise. A filtered image can then be obtained by optical or digital rediffraction of the unobstructed portion of the pattern. A three-dimensional reconstruction can also be done by combining multiple views of the structure in a computer.

Optical Diffraction and Filtering

The intermediate stage in an image formation is the diffraction of radiation by an object, and the apparatus for Fourier transformation serves for describing this phenomenon mathematically. *Optical diffraction* is the simplest, most commonly used initial step for most image-processing studies. It allows an objective assessment and reveals the periodic structural information contained in the image. The geometrical arrangement of subunits of the specimen seen in the diffraction pattern often cannot be observed by direct visualization of electron micrographs. An optical diffractometer is used; the image (the plate or film) is illuminated by a laser beam, and a diffraction pattern is formed by interference of light waves passing through the image. A lens focuses this diffraction pattern (Fourier image of the object) in a plane behind the image where it can be observed or photographed. Optical diffractometers are commercially available, but some microscopists build their own diffractometers.

Optical filtration is the second step in the image reconstruction. The method eliminates noise from the image and provides a clear picture, but the technique is only applicable to periodic objects with translational symmetry. The first important step in this process is to index the diffraction pattern correctly. When a pattern is properly indexed, it is possible to determine the spots that arise from image noise (aperiodic details) and those that come from the periodic structure of the specimen. It is, therefore, necessary to distinguish clearly the noise and signal components. A crystalline specimen usually produces clearly identifiable periodic array of the spots, whereas noise or aperiodicities give rise to spots in all parts of the pattern.

When the diffraction spots are found to be consistent with a given lattice, a filter mask with holes is designed. It is accurately positioned in the diffraction plane for allowing unobstructed diffraction spots at the lattice points to pass through. Thus, most of the diffracted noise spots are blocked out by the opaque regions of the mask. A reconstruction lens, placed behind the mask, is then used to refocus the unobstructed rays to reconstruct the filtered image (inverse Fourier transformation). Thus, the filtered image results from a double diffraction phenomenon. The initial diffraction pattern is obtained in the first step, and in the second stage, the reconstruction lens rediffracts the diffracted rays (Fig. 18-2). The diffractometer must be equipped with a high quality reconstruction lens for filtering the image to minimize image distortions.

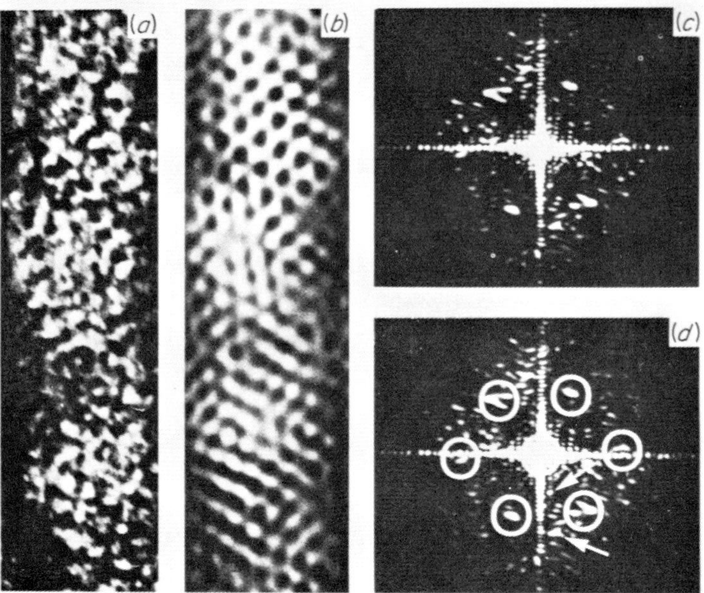

Figure 18-2. Image processing by the optical diffraction and filtering technique. (a) Original micrograph of a wide tube of negatively stained cowpea chlorotic mottle virus. (b) Reconstructed image with noise removed by mask. (c) Diffraction pattern arising from repeating elements and from noise. (d) Diffraction pattern identical to that of (c) but showing the areas of the pattern and for image reconstruction inside seven circles. From J. B. Bancroft, G. J. Hills, and R. Markham, A study of the self assembly process in a small spherical virus. Formation of organized structures from protein subunits *in vitro*, *Virology*, *31*:354, 1967. Courtesy of Academic Press, New York.

Computer Processing of Electron Micrographs

Optical image processing methods have the advantage of being simple, but they are not as versatile as digital image processing techniques. The digital method can provide a quantitative analysis and extreme flexibility in manipulation of data that is not possible with optical methods. In addition to eliminating noise, it is easier digitally to correct image aberrations, e.g. astigmatism, spherical aberration, and defocusing. Three-dimensional reconstruction and rotational filtering are usually impossible by optical Fourier techniques. Individual image reconstructions can be averaged together by digital processing and a quantitative measurement of their agreement assessed. The optical diffraction patterns record only the amplitude of Fourier transformation. The digital computation technique can quantitatively measure amplitude and phases and alter them if necessary for contrast transfer effect. However, the method is complex and expensive. When only a qualitative examination of diffraction pattern is required, optical methods are quick and inexpensive.

The computer enhancement of images originated with its application to space photography. It is a fairly recent addition to the tools at the disposal of electron microscopists, and processing of TEM images of biologic specimens by computer is still in its infancy. Before carrying out digital computation, a number of electron micrographs are usually examined by optical diffraction, and approximately ten images of the best preserved specimens are selected for computer processing.

MICRODENSITOMETRY: The computer must be provided with a discrete digitized version of the micrograph. It is, therefore, necessary to convert the optical density in the image to an array of numbers by densitometry. A photographic negative is scanned by a microdensitometer, which measures the optical density of small grid areas generally 25 to 50 μm apart (some images may require finer sampling areas than this). The maximum number of gray levels is tailored to the properties of the computer, and a common figure is 256. Microdensitometry converts the optical information into electrical information so that the computer is able to calculate the contrast of each grid of the image and ultimately determine the characteristics of the object.

Various models of microdensitometers are commercially available of which the *drum-type scanners*, e.g. Scandig, Photoscan, are the fastest. The micrograph must be recorded on film for this type of scanner. Although the micrograph from a plate can be transferred to a film, this is not recommended because it may introduce contrast changes. Another type of densitometers is the *flatbed type*, which accept both plates and film. They are slower than the drum scanners but may achieve greater precision. Microdensitometers are quite expensive, and unfortunately there is no system at present that can be used to directly couple the TEM to a computer without passing through the intermediate microdensitometry. The microdensitometer measurements can be recorded on a magnetic tape for future processing in the computer or can be sent to the computer.

COMPUTING: An ideal computer for image processing should have a large memory with one or more cathode ray tubes (CRTs) for displaying the image. Computers usually are fast processors of otherwise slower operations, e.g. for Fourier transformation that may have to be repeated many times before the final image is obtained. The digitized image can be written on a magnetic tape and filed in a large computer. Some microscopists, however, have used minicomputers, e.g. PDP 8/E, PDP 11/25, PDP 11/45, for this purpose.

The information provided by contrast data is then mathematically manipulated in the computer by fast Fourier transformation, and the diffraction amplitudes and phases are displayed. The Fourier transformation separates the image structure into its linear or spatial frequency distribution. This allows the complex signals to be described in terms of their simple components, and a mathematically more manipulable equation of the image can be obtained. The diffraction pattern can then be filtered by reverse Fourier transformation and the results displayed by any of a number of ways, e.g. film writing. The forward and reverse Fourier transformation process may have to be repeated many times to correct the amplitude and phase shifts, instrumental aberrations, and defocusing to yield all necessary information to reconstruct the image in the computer. This reconstructed signal is then fed to a CRT, which displays a filtered image of the original specimen. A high frequency filter is commonly used, and filtering by direct convolution is also possible. The microscopist can scrutinize the image as it proceeds and change the course of events when necessary. It is important that the quality of the processed image be at least as good as that of the original. Computer processing has given best results with periodic objects (Fig. 18-3), but the same principle can also be applied to aperiodic objects.

The micrographs of aperiodic objects vary widely, and their processing is usually much more difficult than periodic objects. A considerable experimentation with these images is required for selecting the best processing operations and parameters. A box convolution operation has been applied to these objects in which a box is placed on the image data points and shifted in all possible directions on the image grid. For each position, the average of data points within the box is computed. The resulting image, the *box convoluted image*, has less noise than the original. The processed image contains only the high-resolution information, and fluctuations in background density are removed. Image processing of some aperiodic structures using linear contrast stretching and box convolution followed by histogram equalization in a PDP 11/45 computer with high-pass filter enhancement is shown in Figures 18-4 and 18-5.

Three-Dimensional Reconstruction

A three-dimensional reconstruction of a two-dimensional image is the ultimate goal of the electron microscopist to provide a tangible, lifelike form. The simplest, tiresome procedure consists of serial thin sectioning. The same area in a series of thin sections is examined, and a three-dimensional image is built up by following changes from section to section. Thick sections examined in the HVEM can reduce the time for three-dimensional reconstruction, but the method, nevertheless, is still cumbersome.

A computer-aided three-dimensional reconstruction by Fourier transformation using a series of views through the same direction at different angles is most popular. The number of views required for the reconstruction depends on the symmetry of the object. A helical object, for example, may require only one view, an icosahedral object may need two to four views, but thirty or more views are usually required for asymmetrical objects. The Fourier transformation gives a three-dimensional image by taking into account the symmetry of the object. The tail sheath of the T_4 bacteriophage was reconstructed in this manner (Fig. 18-6). Because of its helical symmetry, only a single two-dimensional micrograph was needed, which effectively contained many different views of the structure.

Other methods of three-dimensional reconstruction use the algebraic relation between the three-dimensional object density and the projected densities. Photographic superimposition methods have also been used for reconstruction. The micrograph or the negative, in this method, is rotated or translated to superimpose the

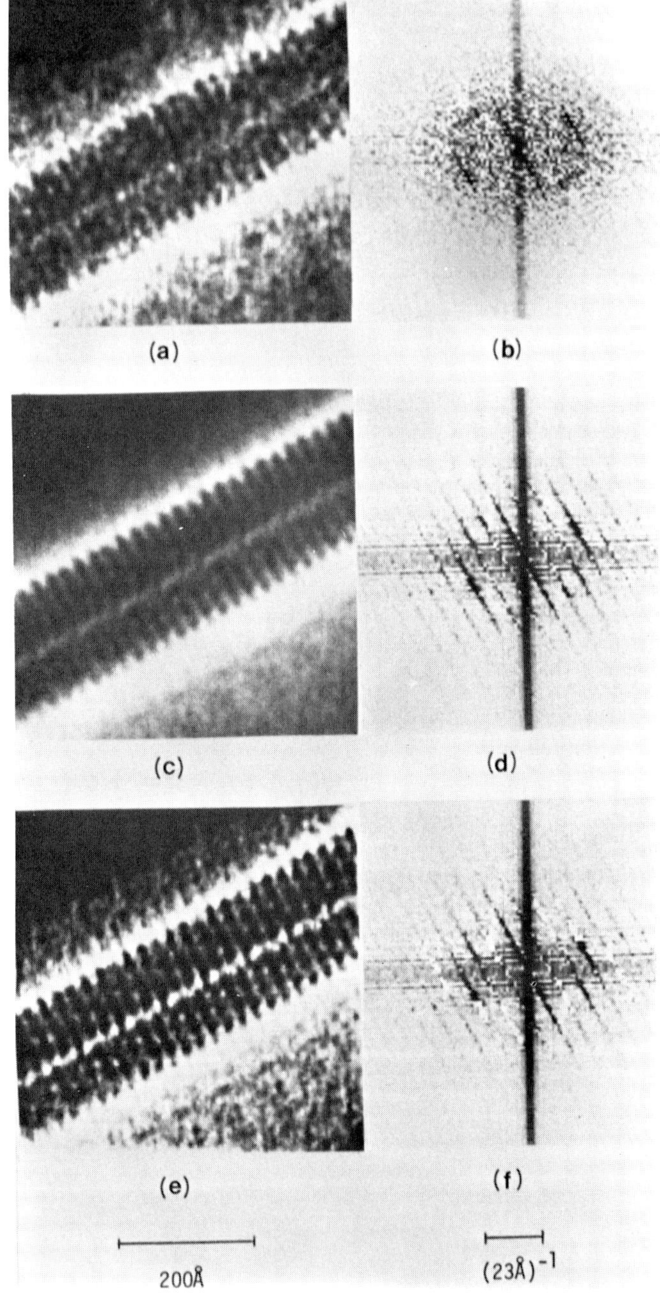
(a)
(b)
(c)
(d)
(e)
(f)
200Å
(23Å)⁻¹

images of individual asymmetric units in the specimen. A composite picture of the symmetry equivalent parts of the specimen is then formed on a single photographic plate that produces an averaged image.

The usefulness of the computer in the STEM (Crewe type) has already been demonstrated. The computer can also be useful in the SEM where the information from different types of images can be recognized and appreciated more easily. It appears that more electron microscopists will use the computer in the future for analyzing the images of biologic specimens.

SELECTED BIBLIOGRAPHY

Baker, T.S.: Image processing of biological specimens: A bibliography. In Griffith, J.D. (Ed.): *Electron Microscopy in Biology*, vol. 1. New York, Wiley, 1981.

Bancroft, J.B., Hills, G.J., and Markham, R.: A study of the self assembly process in a small spherical virus. Formation of organized structures from protein subunits *in vitro*. *Virology, 31*:354, 1967.

Cowley, J.M.: The principles of high-resolution electron microscopy. In Hayat, M.A. (Ed.): *Principles and Techniques of Electron Microscopy: Biological Applications*, vol. 6. New York, Van Nostrand, 1976.

DeRosier, D.J. and Klug, A.: Reconstruction of three-dimensional structures from electron micrographs. *Nature (London), 217*:130, 1968.

Frank, J.: Computer processing of electron micrographs. In Koehler, J.K. (Ed.): *Advanced Techniques in Biological Electron Microscopy*. New York, Springer-Verlag, 1973.

Glaeser, R.M.: Radiation damage and resolution limitations in biological specimens. In Swann, P.R., Humphreys, C.J., and Goringe, M.J. (Ed.): *High Voltage Electron Microscopy*. London, Academic Press, 1974.

←

Figure 18-3. Computer processing of scanning transmission electron microscope (STEM) images of tobacco mosaic virus (TMV). (a) Original micrograph of TMV (intact helix) stained with 1% uranyl acetate. (b) The magnitude of Fourier transform of (a). (c) (a) averaged over nine periods of the helix. (d) The magnitude of Fourier transform of (c). (e) Processed image of one side of TMV helix with visible subunits. (f) The magnitude of Fourier transform of (e). The horizontal and vertical lines through the Fourier transforms are scanning noise. The image was processed in a PDP 11/20 computer applying a high frequency emphasis filter and high frequency cutoff. From E. J. Kirkland, W. T. Freeman, M. Ohtsuki, M. S. Isaacson, and B. M. Siegel, Digital image processing of STEM images of tobacco mosaic virus (TMV), *Proceedings of the Electron Microscopy Society of America, 38th Annual Meeting*, 40, 1980. Courtesy of Claitor's Publishing Division, Baton Rouge, Louisiana.

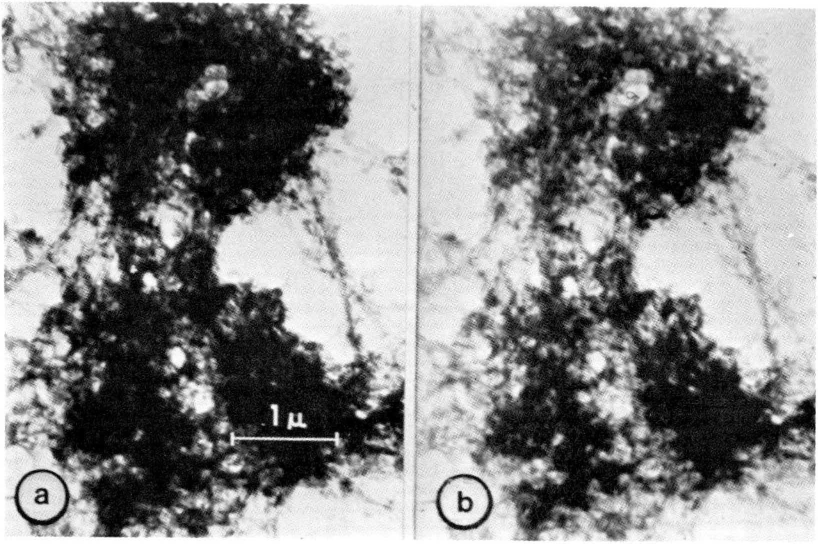

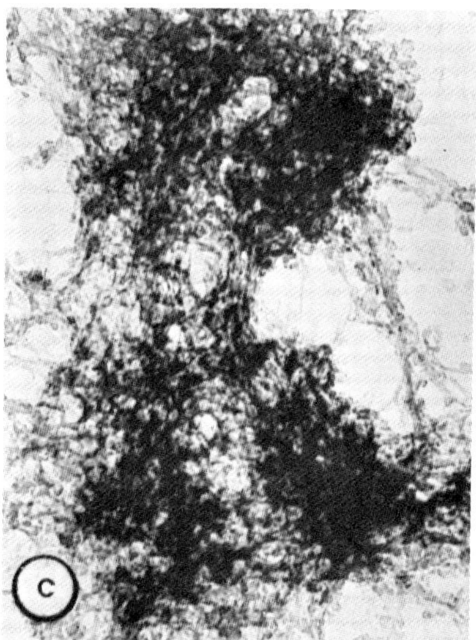

Figure 18-4. Digital image enhancement of Chinese hamster chromosomes. (a) The original micrograph. (b) Contrast stretched with a linear function between the 1% to 99% histogram values. (c) Box convolution (box size 5 × 5) followed by histogram equalization used for high frequency enhancement. From W. Goldfarb and J. Frank, Digital image enhancement for transmission electron microscopy, *Ninth International Congress in Electron Microscopy (Toronto), Vol. III*:22, 1978. Courtesy of the Microsurgical Society of Canada.

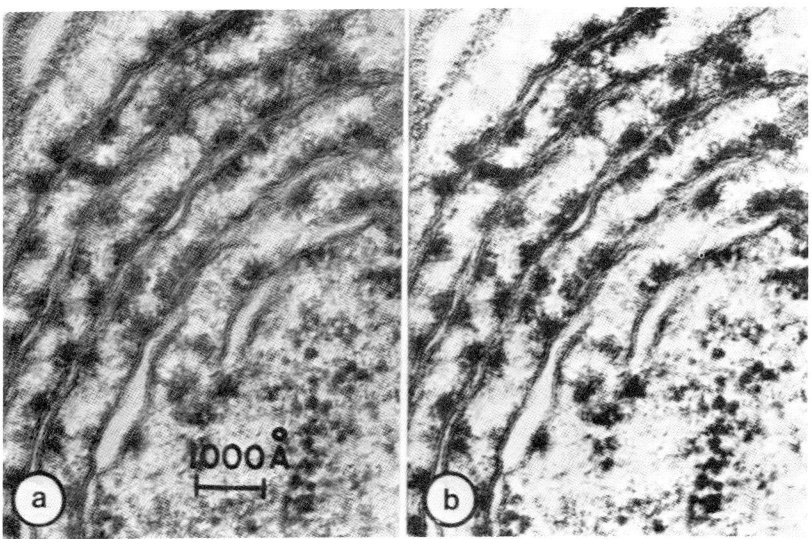

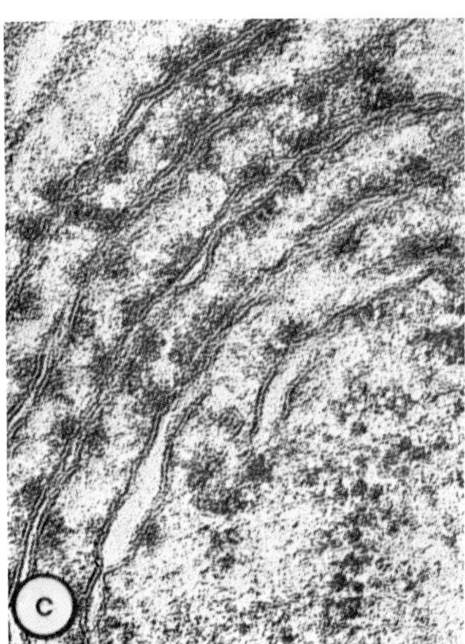

Figure 18-5. Computer image processing of photosynthetic lamellae of the Blue Green algae *Syechococcus lividus* with attached phycobilisomes. (a) Original micrograph. (b) and (c) are the same as Figure 18-4 (b and c). The box-convoluted images contain only the high-resolution information. Therefore, the edges appear enhanced, fluctuations in the background density are removed, but the noise is also enhanced, e.g. see the granularity between the double membrane in (c). From W. Goldfarb and J. Frank, Digital image enhancement for transmission microscopy, *Ninth International Congress in Electron Microscopy (Toronto), Vol. III*:22, 1978. Courtesy of the Microscopical Society of Canada.

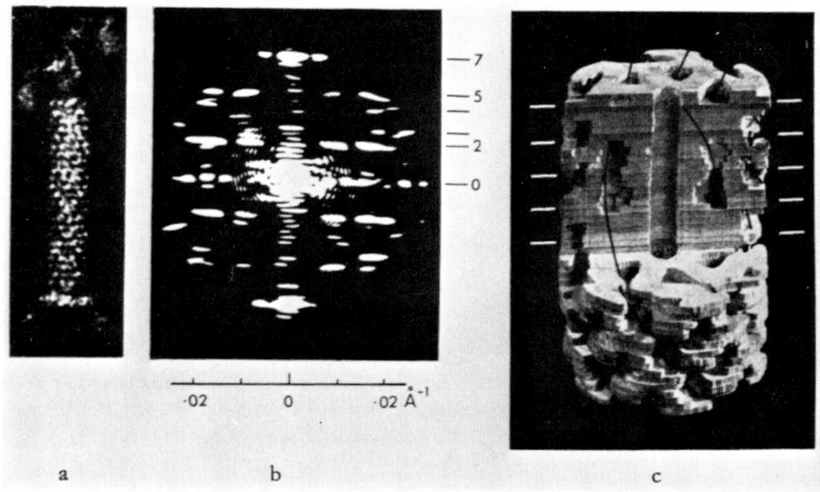

Figure 18-6. Three-dimensional reconstruction of the T_4 bacteriophage tail at a resolution of 15Å. (a) Negatively stained original micrograph. (b) Its optical diffraction pattern. (c) The three-dimensional reconstruction model. From D. J. DeRosier and A. Klug, Reconstruction of three-dimensional structures from electron micrographs. Reprinted by permission from *Nature, 217:*130. Copyright © 1967, Macmillan Journals Ltd.

Goldfarb, W. and Frank, J.: Digital image enhancement for transmission electron microscopy. *Ninth International Electron Microscopy (Toronto), vol. III:*22, 1978.

Hawkes, P.W.: Computer processing of electron micrographs. In Hayat, M.A. (Ed.): *Principles and Techniques of Electron Microscopy: Biological Applications*, vol. 8. New York, Van Nostrand, 1978.

Isaacson, M.S.: Specimen damage in the electron microscope. In Hayat, M.A. (Ed.): *Principles and Techniques of Electron Microscopy: Biological Applications*, vol. 7. New York, Van Nostrand, 1977.

Kirkland, E.J., Freeman, W.T., Ohtsuki, M., Isaacson, M.S., and Siegel, B.M.: Digital image processing of STEM images of tobacco mosaic virus (TMV). *Proc EMSA, 38th Annual Meeting*, 40, 1980.

Klug, A. and Berger, J.E.: An optical method for the analysis of the periodicities in electron micrographs, and some observations on the negative staining. *J Mol Biol, 10:*565, 1964.

Langmore, J.P.: Electron microscopy of atoms. In Hayat, M.A. (Ed.): *Principles and Techniques of Electron Microscopy: Biological Applications*, vol. 9. New York, Van Nostrand, 1978.

Markham, R., Hitchborn, J.H., and Hills, G.J.: Methods for enhancement of

image detail and accentuation of structure in electron microscopy. *Virology, 20:*88, 1963.
Reimer, L.: Review of the radiation damage problem of organic specimens in electron microscopy. *Proc EMSA, 31st Annual Meeting*, 476, 1973.
Welton, T.A.: Computational correction of aberrations in electron microscopy. *Proc EMSA, 29th Annual Meeting*, 94, 1971.
Williams, R.C. and Fisher, H.W.: Electron microscopy with minimal beam damage. *Proc EMSA, 28th Annual Meeting*, 304, 1970.
Zeitler, E.: Electron matter interaction. *Ninth International Congress on Electron Microscopy (Toronto), vol. III:*29, 1978.

Chapter 19

DIAGNOSTIC ELECTRON MICROSCOPY AND FUTURE OF BIOLOGIC ELECTRON MICROSCOPY

THE ELECTRON MICROSCOPE (EM) has become an important tool for medical diagnosis, especially in surgical pathology. Many large hospitals now have at least one EM, and smaller hospitals and private pathology laboratories often use the EM facilities of nearby large hospitals. Although the use of the EM is not yet truly routine and it is not going to replace the light microscope in medical diagnosis, it may be indispensable for final and more dependable diagnostic problemsolving and may complement routine light microscopy.

Light and transmission electron microscopy can be done on the same biopsy material, and ultrathin sections of these tissues are invaluable in making an early diagnosis of diseases. The rapid embedding procedure (*see* Chap. 13) may be applied to biopsy specimens. Biopsies of the small intestine are frequently used as an aid in diagnosis of maladsorption state, e.g. celiac disease. The transmission electron microscope (TEM) has contributed significantly to the understanding of the pathogenesis of renal diseases, particularly those affecting the glomerulus, and its diagnostic value has been clearly demonstrated. The diagnosis of kidney diseases such as lupus nephritis, glomerulonephritis, lipoid nephrosis, and amyloidosis has been confirmed by electron microscopy. The diagnostic value of the TEM in muscle diseases has been reported from muscle biopsies and has been used in studies of the earliest and latent phases of biopsies. Many ultrastructural alterations in skeletal muscle and peripheral nerve biopsy specimens, however, may be nonspecific. Therefore, these findings should be interpreted with caution in combination with clinical data and light microscopic findings.

It has been reported that the diagnosis of curvulinear body disease can only be made by electron microscopy. These bodies are found in several neurodegenerative diseases and are present in urine sediments, muscles, and conjunctival exudates that can only be seen with a TEM. Axonal alterations and demyelination can also be seen by thin-section electron microscopy. A diagnosis of abnormal cilial morphology in a chronic respiratory disease of males has also been reported. Ultrastructural studies may distinguish certain lymphomas or leukemias and, along with light microscopy, subclassify them precisely. Both the TEM and scanning electron microscope (SEM) can be used for early cytodiagnosis. The TEM is now increasingly used for identification of tumors that cannot be classified by light microscopy. Immunologic methods using ferritin-conjugated and enzyme-labeled antibody techniques have also been used for identifying autoimmune diseases such as lupus erythematosus.

The identification of virus particles by the negative staining technique in the TEM is extremely simple, and many viral diseases can be quickly diagnosed by this method. Rota- and coronaviruses cause neonatal diarrhea, and parvoviruses cause diarrhea in adults. These particles are shed in diarrheal stool in large quantities, and a simple diagnosis can be rapidly made by electron microscopy and immune electron microscopy (Chap. 16) of these stool samples. The identification of hepatitis A virus particles in stool has also been done by this method. The TEM is invaluable in the rapid diagnosis of neonatal viral diarrhea in animals. Other applications of viral identification include herpes simplex and pox viruses from vesicular fluids of skin lesions. Virus particles, e.g. cytomegaloviruses, have also been identified in biopsy specimens.

Pathology plays a major role in the forensic science, and the analytical SEM has proved invaluable in this area. Back scattered and secondary electrons along with x-rays have been successfully used for surface as well as elemental analysis. The x-ray microanalysis has been used for detecting the levels of lead in infants' teeth, heavy metals in tissues, and asbestos fibers in the lungs. The advantages of electron microscopic diagnosis outweigh the cost and time involved, and its use will become routine in rapidly diagnosing an increasing number of diseases.

FUTURE OF BIOLOGIC ELECTRON MICROSCOPY

The ultimate goal of the biologist is to image the individual atoms in a biomolecule by electron microscopy. This has already been achieved with heavy metal atoms. However, the fundamental limit to observing single atoms of biomolecules is the radiation damage (*see* Chap. 18), which cannot be prevented. The atoms of biomolecules do not produce enough contrast, and the interatomic distance of ≈ 1.5Å of their lighter elements can hardly be resolved with the modern high-performance TEMs.

High resolution now refers to imaging the distribution of molecules or atoms. Spherical aberration is one of the most important factors in limiting the resolution, and a spherical aberration corrector is presently being constructed for the most sophisticated scanning transmission electron microscope (STEM) of the Crewe type. With the construction of 1 MV STEM in Chicago, the resolving power is predicted to approach 1Å, and the STEM may be the answer to visualize individual atoms of biomolecules. A high-brightness field emission gun, ultrahigh vacuum with ion pumps, low-noise support films made of single graphite crystals, and liquid helium temperature for reducing radiation damage have to be used for imaging these atoms. The sequencing of DNA appears to be very difficult by electron microscopy at present, and the EM has to compete with reliable, rapid biochemical techniques for this purpose.

Historically, it was relatively simple to improve the resolving power of the conventional TEM from 100Å to 10Å in a short span of 8 years (1938-46). Since then, it has become quite difficult to improve the resolving power beyond 2 to 3Å. It will require highly advanced technology to obtain a resolving power of 1Å. It appears, however, that it will be done, and imaging of individual atoms of biologic molecules will be achieved. The record of biologic electron microscopy is good, and the future holds great promise.

SELECTED BIBLIOGRAPHY

Carter, H.W.: Scanning electron microscopy for pathologists. *EMSA Bulletin, 10*:46, 1980.

King, D.W.: *Ultrastructural Aspects of Disease*. New York, Harper & Row, 1966.

Mackay, B.: *Introduction to Diagnostic Electron Microscopy*. New York, Appleton-Century-Crofts, 1981.

Meek, G.A.: *Practical Electron Microscopy for Biologists*, 2nd ed. London, Wiley, 1976.

Watson, J.H.L.: Electron microscopy in medical diagnosis. *Microscope, 17*:63, 1969.

APPENDIX

GENERAL ELECTRON MICROSCOPE SUPPLY SOURCES

Beem, Inc., Box 132, Jerome Avenue Station, Bronx, New York, New York 10468.

Electron Microscopy Sciences (EMS), Box 251, Ft. Washington, Pennsylvania 19034.

E.F. Fullam, Inc., Box 444, Schenectady, New York 12301.

Ladd Research Industries, Inc., Box 901, Burlington, Vermont 05401.

Ted Pella, Inc., Box 510, Tustin, California 92680.

Blazers Union, 8 Sagamore Park Road, Hudson, New Hampshire 03051.

Polysciences, Inc., 400 Valley Road, Warrington, Pennsylvania 18976.

SPI Supplies, Box 342, West Chester, Pennsylvania 19380.

LKB Instruments, Inc., 12221 Parklawn Drive, Rockville, Maryland 20852.

M & W Systems, 39124 State Street, Fremont, California 94538.

Scanning Electron Microscopy, Inc., Box 66507, AMF O'Hare, Illinois 60666.

Efjeld Co., Inc., 3 A Street, Burlington, Massachusetts 01803.

Tousimis Research Corporation, Inc., 2211 Lewis Avenue, Rockville, Maryland 20851.

INDEX

G

H

I

S

T

U

V